AMER
CLOCKS

An Introduction

Tom Spittler

Clocks Magazine Beginner's Guide Series № 3

Published by Splat Publishing Ltd.
141b Lower Granton Road
Edinburgh
EH5 1EX
United Kingdom

www.clocksmagazine.com

American Clocks, An Introduction is based on a series of articles which originally appeared in *Clocks* magazine

ISBN: 978-0-9562732-2-2

2 4 6 8 10 9 7 5 3 1

Printed by Stephens & George, Goat Mill Road, Dowlais, Merthyr Tydfil CF48 3TD, U.K.

CONTENTS

Clocks Magazine Beginner's Guide series

No. 1. Clock Repair, A Beginner's Guide
No. 2. Beginner's Guide to Pocket Watches
No. 3. American Clocks, an Introduction

In preparation

No. 4. Collecting Pocket Watches, 1750-1920
No. 5. Make Your Own Clock, A Beginner's Guide
No. 6. Beginner's Guide to Black Forest Clocks

Preface & Acknowledgements

I had two goals in mind in undertaking the task of compiling and writing this book, and those were to cover the subject of American clocks thoroughly, and to make the subject easily understood by someone with no or limited knowledge of clocks in general or, specifically, American clocks.

It is impossible to understand American clocks without understanding the country itself, its manufacturing methods, transportation, capitalization, communications, free enterprise and, most importantly, its people. All these factors are woven into the fabric of America … and American clocks.

In writing this book, I have tried to keep the use of technical terms to a minimum. Those that are used are explained in the Glossary on page 125. For how a clock works, see Appendix 1 on page 121.

A book such as this requires the efforts of numerous people and I thank all of those who have assisted me in any way. Foremost, I thank Dr. Snowden Taylor, who never failed to "educate" me whenever I asked. He was untiring and I could never have completed the task without his assistance.

I would also like to thank Chappell Jordan Clock Galleries of Houston, Texas; Cowan's Auctions, Inc., of Cincinnati, Ohio; Delaney Antiques of West Townsend, Massachusetts; and Skinner, Inc., of Boston, Massachusetts, for all the wonderful photographs of clocks they provided.

I thank Chris Bailey, Bill and Rusty Bergman, Terry Brotherton, Chris Brown, Freda Conner, Robert Cheney, Doug Cowan, David and Brent Cox, Tom and Fran Davidson, Diana DeLucca, Paul Foley, George and Cathy Goolsby, Earl Harlamert, Carter Harris, Roger Huegel, Terry and Vanda Lawrence, Greg and Paul McCreight, Chris Peck and Ralph Pokluda, who all assisted me greatly, again mostly with photographs and/or access to clocks, and Doug Stevenson for his encouragement.

Finally, I would like to thank John Hunter, Editor of *Clocks* magazine, for giving me the opportunity and encouragement to write this book.

Tom Spittler, Ohio, 2011

CHAPTER 1
AMERICAN CLOCKS, COLONIAL TO 1900

Figure 1.1 (above). Eli Terry (1772-1852), Plymouth, Connecticut, famous American clockmaker/industrialist.

Figure 1.2. American 30-hour wooden clock movement, about one-third larger than brass grandfather movement. Photograph courtesy Cowan's Auctions, Inc., Cincinnati, Ohio.

To understand American clocks it is important to understand each wave of American clock type that broke over the country, as well as the storm that generated that wave. Unlike Britain and Europe, where longcase, table and wall clocks lived side by side for many years, in America each new clock type drove the previous favorite from the market. This book will explain the major reasons this occurred. It will also help the reader slot each type of clock into the period it was produced.

The first (and most important) factor affecting American clockmaking and clocks is the country's rapid move from the colonial system to the "freedom" allowed by the new nation in the late 1700s. The second factor is that Americans couldn't make springs. The final factor concerns what we will call the "unintended consequences of mass production."

The settlement of the original American colonies was largely done by the British, and therefore it is the British eighteenth century colonial system that must be understood in order to understand early American clockmaking. The colonies were supposed to develop the land and send raw materials and excess agricultural products to Britain to be developed into the manufactured goods needed by her and her colonies. Military protection would be provided

Figure 1.3. A Joseph Ives pillar-&-scroll shelf clock with turned feet, a late feature, ca. 1828. Photograph courtesy Skinner, Inc., Boston, USA.

Figure 1.4. Very rare wooden Ives movement with "wagon-spring" power. Photograph courtesy Cowan's Auctions, Inc., Cincinnati, Ohio.

Factors Affecting American Clockmaking

The "freedom" allowed by the new nation
Inability to make springs
Unintended consequences of mass production:

- **Local market saturation**
- **Large amount of capital required**
- **Fragile clocks don't ship well**

by Britain, assisted by the colonies. The colonists would build their own homes and could manufacture any of the simple household goods—chairs, tables, etc.—they required. They would obtain more complex manufactured items from Britain.

Like most systems, things never worked the way they should have. In fact, where it was advantageous to do so, such as with the sale of cotton, lumber, tobacco and indigo dye to Britain, the 13 colonies were all too glad to participate. When it came to the colonial rich purchasing fine British manufactured goods, including clocks and watches, they were also pleased to do so. However, when items that should have been manufactured in Britain could be made in the colonies and the skills and materials existed to do so, the colonists would make them and were not challenged.

Skilled workmen arrived from Britain, men such as clockmakers Peter Stretch, who came to Philadelphia in 1702, and James Batterson, who came to Boston in 1707 by way of Philadelphia. The arriving British clockmakers almost immediately advertised for "smart young lads" to become apprentices and soon there were enough American-born clockmakers to take care of the needs of the Americans, all well trained through the same apprentice system that

Figure 1.5. Marsh hollow-column clock. The hidden weights run in hollow columns at side. Photograph courtesy Cowan's Auctions, Inc., Cincinnati, Ohio.

Figure 1.6. Pillar-&-scroll clock, ca. 1825, an invention of Eli Terry. Photograph courtesy Cowan's Auctions, Inc., Cincinnati, Ohio.

existed in Britain.

That British apprentice system protected the clockmakers in two ways. First, it ensured that young apprentices whose parents had paid a huge sum—about a quarter of a year's income—were fully trained as clockmakers, apprised of all the secrets of the trade. Second, it ensured that once they were trained and established within a city or town, that city or town would not allow excess clockmakers "freedom" to practice their trade. It also protected the public by ensuring they were getting a quality clock made by a properly trained clockmaker who could service it when necessary.

This was the system in 1770. Then came the Revolution with its cries for liberty and freedom. After an eight-year war the Americans won their liberty and freedom, and to many this meant an end to the restrictions to practice any trade. An individual would not be prohibited from practicing a trade, regardless of the skills required, and that individual could practice it wherever he chose. The clockmaker without skills would be allowed to fail if his products weren't of good quality and no system would be in place to protect the customer from his shoddy goods. Likewise there was no system to come to the clockmaker's aid when he failed. That was the consequence of liberty and freedom.

Some cities such as Philadelphia did establish clockmakers' guilds to protect trade but had difficulty enforcing any restrictions on those who did not join. Philadelphia was only successful as long as the major supplier of clock and watch material, John Wood, would not sell to anyone within the city that was not a guild member.

There was, however, an even more sinister aspect of this new freedom to trade without restrictions. No longer were young workers protected. Young lads of little or no means would

Figure 1.7. Carved column-&-splat clock showing American Empire period influence, ca. 1832. Photograph courtesy Cowan's Auctions, Inc., Cincinnati, Ohio.

Figure 1.8. Fully developed half-column-&-splat clock, ca. 1835, without feet. Photograph courtesy Cowan's Auctions, Inc., Cincinnati, Ohio.

find themselves serving apprenticeships to masters who would only teach them one aspect of the trade. With no limit on the number, a master could have many apprentices, each performing one of the operations—such as cutting wheel teeth—required to make a clock. These lads would never learn a complete trade and would never be able to become independent clockmakers. They were trapped in a system much like slavery except they were free to leave if they wanted to after they served their time.

One of the first clockmakers to take advantage of this new system was Thomas Harland, a British clockmaker born in 1735 and working in Norwich, Connecticut, by 1773. He claimed in December 1773 that he learned the clock and watch trade in London, but local sources say he spoke with a Scottish accent and was known as the "Old Scotsman." He "trained" over 21 apprentices and had from 10 to 12 in his workshop at one time. According to Chris Bailey, Curator of the American Clock & Watch Museum of Bristol, Connecticut, the Harland clocks he has examined showed a very wide range in quality of workmanship from very good to very poor, some among the worst he's seen. Many of Harland's apprentices, however, went on to have great careers and produce excellent clocks. One in particular, Daniel Burnap, kept his apprentice journal which has been published. Dr. John Robey, author of *The Longcase Clock Reference Book*, says it is the best source of information on how longcase movements were made and, in fact, it is the only source he is aware of that gives step-by-step details of the process.

Daniel Burnap, in turn, trained Eli Terry, figure 1.1, who was famous for being the first individual to use mass production to produce complex manufactured items—a wooden tall clock movement, figure 1.2. (From this point, the term "tall clock" will be used to describe a

Figure 1.9. Triple decker Empire shelf clock. Eight-day brass movement, ca. 1830s. Photograph courtesy Cowan's Auctions, Inc., Cincinnati, Ohio.

Figure 1.10. Chauncey Jerome OG clock with 30-hour rolled-brass movement. Photograph courtesy Cowan's Auctions, Inc., Cincinnati, Ohio.

grandfather clock made in America.)

The second factor that had a major bearing on American clockmaking, as mentioned above, is the fact that Americans could not produce coiled springs. When Americans made clocks of a type where the British and the Europeans would have used springs, they used weights or, in just a few examples, they imported the very expensive springs. The Americans were not successful in making steel coil springs until about 1850, but they mastered the art of making inferior brass springs before 1840. Any American clocks with those brass springs are considered collectible and the springs should not be replaced with modern steel ones unless broken. Even then the broken brass spring should be retained with the clock.

Prior to mastering the manufacture of coiled springs, Joseph Ives successfully manufactured "wagon-spring" clocks starting in the late 1820s. These used for their power source a leaf spring similar in appearance to those used in the suspensions of horse-drawn carts or wagons. His first such clock was a late pillar-&-scroll clock with turned feet, the feature that gives away its c1828 date, figures 1.3 and 1.4. Later, in the 1840s, Ives allowed Birge & Fuller to manufacture many wagon-spring clocks. In the mid 1850s he also allowed another firm, Atkins, Whiting & Co., to make wagon-spring clocks, some of which ran 30 days on only a few inches of travel of a very powerful, multi-leaf wagon spring.

These wagon-spring clocks never proved successful—although they are very collectible. Other American clocks were made in imitation of French portico clocks. Instead of the springs used in the French clocks, the Americans made these clocks with hidden weights running down hollow columns, figure 1.5. It wasn't until the 1850s that the Americans were able to successfully manufacture inexpensive quality steel coil springs.

The last factor influencing why American clocks were the way they were was the "unintended consequences of mass production." As mentioned earlier, Eli Terry was the first to manufacture

Figure 1.11. Beehive clocks, both Brewster & Ingrahams with rack & snail striking, date 1840s. Clock left, with springs in cast-iron frame, estimate $4,000. Clock right estimate $400. Photograph courtesy Skinner, Inc., Boston, Massachusetts.

any complicated item using the principles of mass production, namely making parts to such close tolerances that they were interchangeable. Simple chairs had been made in large batches where, through the division of labor, standardized parts such as seats, backs and rungs were produced and then assembled by other workers with no woodworking skills. But no item as complex as a clock, with its requirements for very close tolerances, had been mass produced.

Before Terry took on the mass production of wooden tall clock movements in runs of several thousand, he produced these movements with interchangeable parts in batches of about 25. He then found the first unintended consequence of mass production (or even large batch production). It was that the produced items were all in one place and quickly saturated the ready market. Terry found that he spent more time in traveling further and further from his home in Connecticut to sell the clocks (the clocks did not include a case, only the movement, dial, pendulum, weights and hands) and that traveling and selling was taking more of his time than production. With each batch, that problem became worse, requiring travel further and further from home to find fresh customers.

The second problem he faced was that a large amount of capital was required to tool his factory for mass production, maybe even a whole new factory. He did not have this capital.

The solution came from two individuals, unrelated, both named Porter. They were captivated by Terry's idea to mass produce the clocks and had the capital to help him, agreeing to pay $4 for each movement and for a total of 4,000 movements to be made in three years. Some of the money was to be paid up front. Terry realized he was giving up some of the potential profit on each clock he produced to the Porters, but he had no other option. The Porters had an army of peddlers who traveled by wagon throughout New England selling many items, mostly tinware. This agreement, which became known as "The Porter Contract", was signed in the summer of 1806.

Figure 1.12. Standard size steeple clocks using three different finial types, ca. 1850s. Photograph courtesy Skinner, Inc., Boston, Massachusetts.

Terry then sold his 20 foot existing factory with water power to his former apprentice and now partner Heman Clark in 1807. He purchased another factory which he equipped with water power. He hired two 21-year-old joiners, Seth Thomas and Silas Hoadley, to assist him. By the end of 1807 the factory was complete—with custom-built machinery they had made—but no clocks had been produced. About 1,000 movements were manufactured in 1808 and the remainder in 1809. Terry's venture was a success and mass production was proven in America. Terry, however, had even grander plans, and as a result he sold this factory to his employees Thomas and Hoadley in 1810 and went out to build another factory to make another clock.

Terry realized that he was giving up a large portion of his potential profit by selling a clock movement without a case. That problem was never going to be solved if he continued to manufacture tallcase movements. Terry's idea was to produce a complete clock with a case not too much larger than the hood of a tall clock. He would use his ideas of interchangeable parts and mass production to make the case as well as the movement. Doing this he would still have to sell his clocks wholesale to peddlers, but would eliminate the cabinetmaker's share of the profit. This clock with a wooden movement would have to run the same 30 hours as the tallcase movement with only a fraction of the drop for the weights. The clock was, in my opinion, the most beautiful clock ever made, the pillar-&-scroll, figure 1.6.

The pillar-&-scroll took a long time in development but was on the market from the late 1810s to the late 1820s. During this period the tallcase clock in all its forms went into rapid decline before it almost completely disappeared in America by about 1840.

There was, however, a problem with the pillar-&-scroll. Beautiful as it was, it was much too fragile for miles and miles of travel in a wagon over very crude roads. It was also difficult to pack. This would be the third unintended consequence of mass-production; if the product had

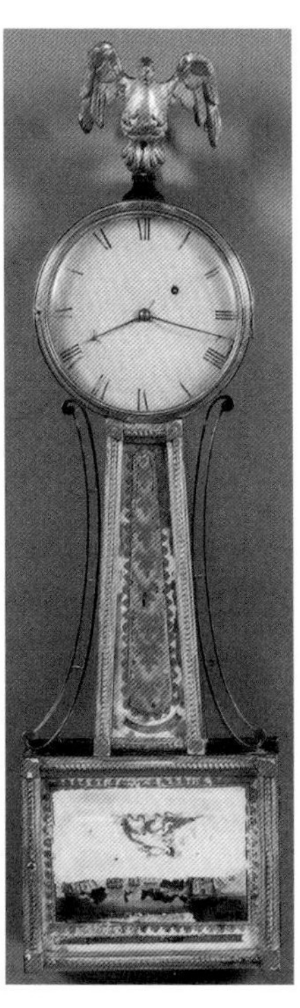

Figure 1.13. Banjo clock by Aaron Willard, ca. 1815. Photograph courtesy Skinner, Inc., Boston, Massachusetts.

Figure 1.14. Gingerbread clock, c1890s. Photograph courtesy Skinner, Inc., Boston, Massachusetts.

to be moved long distances over bad roads it should pack easily and be able to stand up to the journey.

Almost before his development work was done on the pillar-&-scroll, Terry sold Seth Thomas the rights to produce the clock and in the end Seth Thomas probably manufactured more of the pillar-&-scroll clocks than Terry.

During the decade of the pillar-&-scroll, several small firms arose and produced the clock in opposition to Terry's patent. Terry never seriously set about controlling the market and in the end many firms made his clock.

The next clock, an evolutionary development of the pillar-&-scroll, was not an invention of Terry's. It occurred gradually between 1825 and 1830. The elegant but frail feet of the pillar-and-scroll were first replaced with turned or carved feet and then no feet at all in the final form. The fragile freestanding pillars of the pillar-&-scroll were replaced with larger half columns glued securely to the case. The beautiful but very fragile scroll top was replaced by a solid splat at the top of the clock. This was the wooden-movement, 30-hour, half-column-&-splat shelf clock of the 1830s, figures 1.7 and 1.8. Parallel with this clock was an eight-day, upmarket, brass-movement shelf clock which shared popularity with the 30-hour, downmarket, wooden movement, half-column-&-splat shelf clock during the 1830s, figure 1.9.

The first major depression in America occurred in the mid 1830s. That depression took down almost all the clock companies and only about five to ten percent ever returned. The next clock to appear led the recovery. It was the OG, so called because of the S-shaped, ogive or "ogee" curve of the wooden door surround. This used Chauncey Jerome's brother Noble's invention, the 30-hour, weight-driven, rolled-brass movement. That movement, quickly modified to both eight-day and spring drive, was the basis for the majority of the movements for American clocks

Figure 1.15. Walnut parlor clock, c1880s. Photograph courtesy Skinner, Inc., Boston, Massachusetts.

Figure 1.16 (opposite page). Black mantel clock, ca. 1900. Photograph courtesy Vanda Lawrence, New Carlisle, Ohio.

to the end of the nineteenth century. The OG case was the ultimate in ruggedness and ease of packing, figure 1.10.

Transport and sale of clocks out of wagons reduced greatly during the 1840s. Rough roads gave way to canals in the 1820s and 1830s and to the railroads in the 1840s and after. Lines of communication improved greatly with the coming of the US Post Office and the telegraph in the 1840s. The clock companies in Connecticut, as well as sales offices in other major cities, began printing catalogs in the 1850s and the factories could ship small orders directly by rail to small stores all over the USA. Clock markets opened up all over the world. Clocks became smaller with the beehive, steeple and cottage clock leading the way, figures 1.11 and 1.12. Shipping had improved and clocks with frills on their cases began to reappear.

From the late 1830s up to and after the American Civil War of the early 1860s, clock factories became larger and larger and then a second major depression in the mid 1850s concentrated the number of firms into seven major Connecticut firms. That number lasted until the end of the century. The market split into a major domestic (house) clock market and a small but profitable public clock market. Boston, Massachusetts was always a side-center of American clock production starting with the hand-made banjo wall clock, figure 1.13. Boston also concentrated on wall regulators and tower or turret clocks. In numbers, their clocks never competed with nearby Connecticut's domestic clocks, but their clocks were much more expensive and of better quality and no doubt were sold on a much higher profit basis. They relied upon a skilled workforce whereas the Connecticut factories used unskilled labor, men and women with no specific skills and no hope of ever achieving any.

The nineteenth century in America was a time of westward expansion. The population of Connecticut and Massachusetts combined was about three quarters of a million in 1800. In 1850 it was still about the same three quarters of a million, but by way of comparison, a Midwestern state, Ohio, grew from almost no-one in 1800 to 2.3 million by 1850. Immigrants from Ireland and Germany were arriving by the hundreds of thousands a year by the late 1840s. More and more it was the unskilled immigrants, both men and women, eager to work for any wage, who filled the clock factories. However, their American-born children often went west when they came of age. This situation existed for the next 100 years, until well after the demise of the American clock industry.

The major clocks of the last decade of the nineteenth century were the oak gingerbread

clock, along with its upgraded walnut brother the parlor clock, and the black mantel clock, figures 1.14, 1.15 and 1.16. It wasn't until after 1900 that the tambour clock came into fashion. One of these late clocks existed in every sod home on the American prairie. They were truly "everyman's clock." Even the immigrant worker in the clock factory could afford one and even the big white homes on Main Street throughout America had one—usually a gingerbread in the kitchen—thus the name kitchen clock.

But by now there was an even cheaper clock that was out-producing these utilitarian clocks, the ultimate utilitarian—the alarm clock. Alarms were often added, almost as an afterthought, to wood movement shelf clocks in the 1830s and the small shelf clocks with brass spring movements of the 1840s and 1850s. But by the start of the twentieth century, the alarm clock industry was centered in the Midwest near Chicago, Illinois. While great in numbers, these alarm clocks were mostly throwaway items and never seem to have attracted a large number of collectors. They are for the most part ignored today.

By 1900 and the end of the scope of this chapter, the gingerbread and the parlor along with the black mantel clock were the major domestic Connecticut clocks still in production along with the OG in very small numbers. There were a number of schoolhouse and rectangular wall clocks for shops and offices. The railroads ran on regulators from Boston and workers awoke to alarms from Chicago. A few massive hall clocks made their reappearance for the homes of the very wealthy and the offices of senior managers. Hall clocks and regulators aside, the bulk of these 1900 clocks seem to be the focus of today's starting collectors, and as when they were made, they still are the least expensive. But then, everyone has to start collecting somewhere and the American OG, the clock with one of the longest runs of production, has remained one of my favorite clocks for almost 40 years.

Figure 2.1. Philadelphia tall clock in American walnut by Peter Stretch, one of America's first clockmakers. This example dates to about 1730-1735 and stands 8 feet 9 inches tall. Photograph courtesy of Chappell Jordan Clocks, Houston, Texas.

Figure 2.2 (above). Hood and dial of Peter Stretch clock. Stretch arrived in Philadelphia from Leek, England in 1702. Photograph courtesy of Chappell Jordan Clocks, Houston, Texas.

CHAPTER 2
THE BRASS MOVEMENT TALL CLOCK

The King of American clocks, the brass movement grandfather clock, did not have a terribly long reign. Even though American grandfather clocks are known shortly after 1700, an American grandfather from 1750 is a very early example and I have only touched one clock earlier than that date, figures 2.1 and 2.2. When the shelf clock arrived in numbers after 1820, the brass-movement grandfather clock was already in a steep decline brought on by strong competition from the mass produced wooden-movement tall clocks starting in 1810. By 1850 the King was dead. The major period of the American brass-movement grandfather clock was from 1760 to 1840: some 80 years.

A few words about the American terms used to describe these clocks in the period. The terms used were most commonly only "eight-day clock" and "30-hour clock," or just plain "clock." Other than that, "house clock" and "common clock" were sometimes used. "Longcase clock" and "tall clock" are twentieth century terms, while "grandfather clock" came into use in the last quarter of the nineteenth century. Today "tall clock" (or tallcase clock) is used to describe American period grandfather clocks. The adjective "brass-movement" is used in America as some American grandfather clocks had wooden movements; they will be the subject of a later chapter. Often American clocks had "faces" while British clocks had "dials." We will use "dials."

Clockmaking in America was first centered in Philadelphia, Pennsylvania, and Boston, Massachusetts, with German activity appearing by 1750 in towns like York, Germantown and Lancaster in Pennsylvania. The German influence on Pennsylvania clockmaking would last 100 years.

Clockmaking spread out during the eighteenth century from Boston and Philadelphia to nearby areas and bordering states. The clocks from New England (the extreme north eastern states of Massachusetts, Vermont, New Hampshire, Rhode Island and Connecticut, as well as Maine, which was once a part of Massachusetts) shared a common form. They were smaller, more delicate clocks and had similar movements as well as similar painted dials. The clocks made to the south and west of Pennsylvania likewise had characteristics in common with clocks from Pennsylvania. The two states of New Jersey and New York are between New England and Pennsylvania and they developed somewhat similar forms to each other, but different from either New England or Pennsylvania. The above is a very broad generality and, especially when it comes to New York and New Jersey clocks, there are numerous exceptions, figures 2.3, 2.4 and 2.5.

Taking grandfather clocks as a composite of a dial, movement and case, each of these

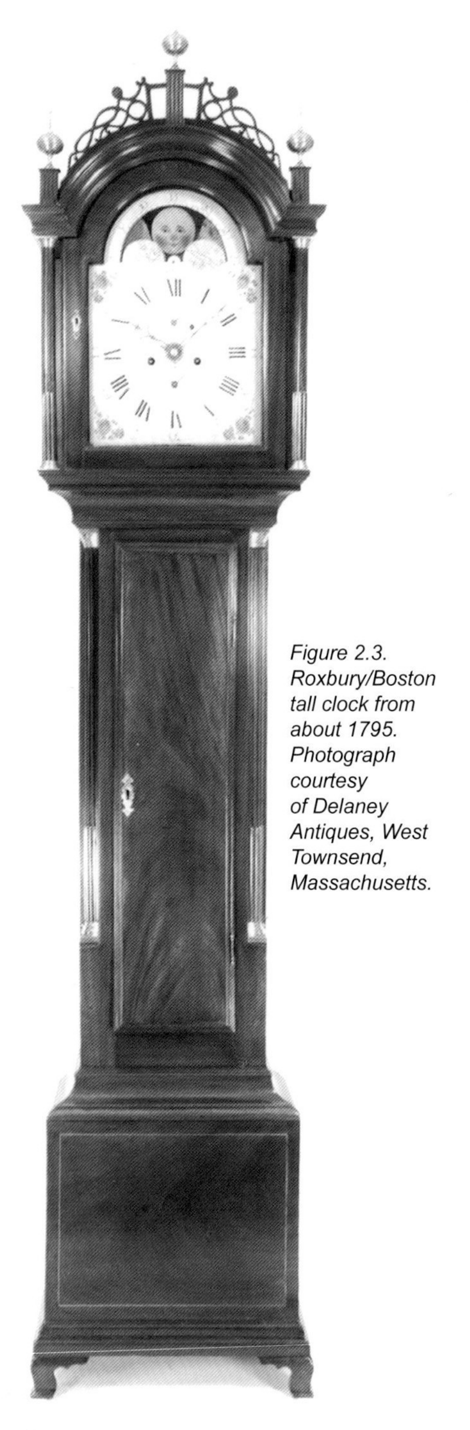

Figure 2.3. Roxbury/Boston tall clock from about 1795. Photograph courtesy of Delaney Antiques, West Townsend, Massachusetts.

components became identifiable as American.

The grandfather clock case was the first part to be identifiable as "made in America." Like European cabinetmakers, American cabinetmakers—sometimes called "furniture makers" in America—made grandfather clock cases while the clockmakers made the movements and sometimes the engraved brass dials. The cabinetmakers used local woods and imported mahogany to make clock cases: walnut, cherry (a large forest tree with pea-size cherries), mahogany, and figured maple as primary woods and pine, poplar and chestnut as secondary woods. If the case was to be painted, which they often were, pine, birch, poplar and plain maple were used. Mahogany from the Caribbean was used for "top-end" clocks, often as a veneer. Note that oak and elm were not used, neither as a primary nor secondary wood.

The movements of American and British clocks are nearly identical except in Pennsylvania where probably half of the eighteenth century clockmakers were German or German-American. The eight-day brass movements these German-American clockmakers made used lantern pinions and a strike-control for rack-striking that had both the rack lifting piece and the rack hook pivoting on the right as viewed from the front, figures 2.6 and 2.7. English-speaking clockmakers used the British system with the rack hook on the left in eight-day examples.

From Boston, clockmaking spread out in the later eighteenth century to the New

Figure 2.4 (opposite page). Chippendale tall clock in mahogany by Joseph Ellicott from Buckingham, Pennsylvania, in 1773. The front of the front plate is signed "1773 #6 Lord" for James Lord, his apprentice, and "Joseph Ellicott 1773 #66 Buckingham, Penn." is on the dial. The Ellicotts were Quakers from Devon, England. Photograph courtesy of Chappell Jordan Clocks, Houston, Texas.

Figure 2.5 (far right). Hood and dial of Joseph Ellicott clock. Photograph courtesy of Chappell Jordan Clocks, Houston, Texas.

Figure 2.6. Eight-day Pennsylvania-German movement from about 1780 with typical strike control having both rack lifting piece and rack hook pivoting from right side of movement. This poor old movement has had many conversions in its lifetime.

Figure 2.7. Open-ended lantern pinion on third wheel of run train of the eight-day Pennsylvania-German movement. Note the reflection in the front plate. You can see that there is only one cap and the trundles or pins just stick out into space, open-ended.

England states, while from Philadelphia clockmaking spread, firstly, to the rest of Pennsylvania and nearby states such as Delaware and Maryland, and then later to Virginia and the western states such as Kentucky and Ohio. Southern states had brass-movement clockmakers, but not many. Some of them migrated down from the north. The exception would be Charleston, South Carolina, where clockmaking flourished during the Colonial period, pre-1776, but decreased gradually thereafter.

New York and New Jersey developed their own style of clocks and some makers used unique movement features, figures 2.14, 2.15 and 2.16. New York City was a large importer of British and European clocks during all periods and local clockmakers never flourished there to the degree they did in other American cities such as Boston and Philadelphia.

The Revolutionary War, 1775-1783, probably caused a pause in clockmaking, but I have not seen any direct evidence of this. The war surely caused a delay to the introduction of the painted dial until about 10 years after it was first seen in Britain. As a bit of irony, it was the American patriot Paul Revere who first advertised these English dials for sale in America in 1785, figure 2.8.

For the next 10 to 12 years all the painted dials in America were imported from England

IMPORTED AND TO BE
SOLD, BY PAUL REVERE,
DIRECTLY OPPOSITE
LIBERTY-POLE, NO. 2, NEW-
BURY-STREET, A LARGE &
VALUABLE ASSORTMENT
OF HARD-WARE GOODS, ...
CLOCK, WATCHMAKERS,
GOLDSMITH & JEWEL-
LERS FILES, TOOLS, &C. ...
ENAMELLED CLOCK FACES,
... WILLARD'S PATENT
CLOCK JACKS, &C.

Figure 2.8. Reproduction of Paul Revere advertisement of 1785.

Figure 2.9. 30-hour movement of Peter Schutz clock from York, Pennsylvania. The dial is dated July 1752. The back or pendulum cock has silk-thread suspension. The silk is missing.

and the painted dials on American clocks were identical to British painted dials. John Ritto Penniman of Boston was among the first dial painters in America and his work was first seen in 1794 when he was just nine years old! In about 1805 a partnership was formed between Boston artist Spencer Nolen and Boston clockmaker Aaron Willard Sr., whereby Spencer Nolen painted dials and Aaron Willard probably provided financial backing. In late 1806 that partnership dissolved and Nolen formed a partnership with Samuel Curtis, another fine artist. The firm of Nolen & Curtis was the first major producer of dials in America. Another early dial production center, but not as large, was Reading, Pennsylvania, and a few of their dials were produced as early as the late 1790s. When the grandfather clock was in steep decline in most of the United States, it hung on for a couple of decades in Pennsylvania and a few other places. Dials for these late clocks were painted by William Jones of Philadelphia, working from 1825 to the early 1840s. There is a high survival rate for these later clocks and many Jones dials are seen. If a clock has an American dial, it was used by an American clockmaker and is on an American clock as none of the American dials were exported.

Movements in America progressed from being completely hand-made by the clockmaker prior to 1795, to taking advantage of kits of brass castings and forged ironwork which became

Peter Schütz
VHRMACHER
YORK

Figure 2.10 (opposite). The dial of the Peter Schutz clock is iron with pewter chapter ring, pewter name boss, and pewter spandrels. The name boss reads: "Peter Schutz / Clockmaker / York / July 52."

Figure 2.11. Later Pennsylvania-German 30-hour rack-striking movement from about 1815.

available to clockmakers just prior to the start of the nineteenth century. Some of these castings and forgings were made in America but most were imported from Britain. A movement made from a set of British castings and forgings would look the same whether it was finished in America or Britain. After 1800, therefore, the look of many American makers' movements changed almost overnight to look as if they were made in Britain. Complete movements were also shipped from Britain to America, albeit in very small numbers.

American brass grandfather movements have vastly different characteristics from place to place and period to period. It would be beyond the scope of this chapter to try to cover all the variations that are known and documentation in this area is almost non-existent. However, a broad overview will follow.

The movements from Pennsylvania can be divided into two types, those that exhibit strong

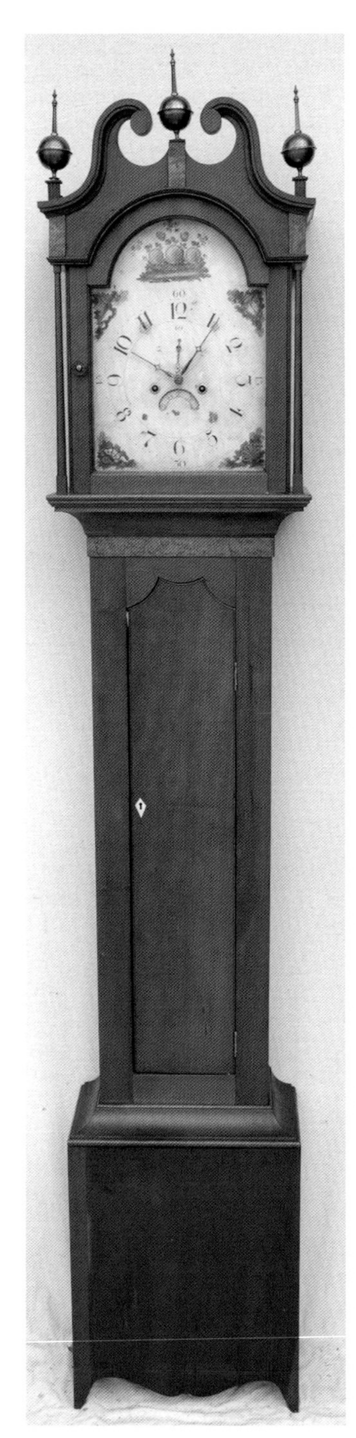

Figure 2.12 (left). New York State tall clock from about 1825.

Figure 2.13 (above). Hood and American dial of New York State tall clock.

Figures 2.14 and 2.15 (opposite page). Skeletonized plates of New York State tall clock. Note large bell and tall bell-stand, and coiled, rather than flat, hammer spring of New York State clock.

Germanic characteristics and those that look British. Starting in about the 1750s, 30-hour Pennsylvania movements made by clockmakers from Germany and Switzerland often have the pendulum suspended with a silk thread and the back or pendulum cock made with two right-angle bends in the vertical plane, figure 2.9. These rack-striking movements are very small and have lantern pinions, sometimes open-ended. They usually have three or four square iron movement pillars between the plates that have round extensions through the front plate all the way to the dial. The dial is screwed from the front to these round extensions to secure it to the movement. The dial is painted iron with pewter spandrels, chapter ring and often a round pewter name boss, figure 2.10. During the next 50 years, these 30-hour German-Pennsylvania movements slowly changed into larger English-looking movements but still retained some of the German characteristics, figure 2.11. "English-looking" movements have the pendulum suspended with a suspension spring, holes in the front plate to accept dial feet, rack striking with the rack hook on the left side in eight-day clocks, cut pinions, brass plates, round plate pillars and no pewter chapter ring.

Though 30-hour clocks exist, they comprise a much smaller fraction of all grandfather clocks in America than in Britain. When they are found they are often from Pennsylvania and fall into this German-American group. Eight-day clocks from Pennsylvania also often exhibit German-American characteristics, the most notable being that the rack striking is controlled by the rack lifting piece and the rack hook both mounts and pivots from the right side of the front plate. Pennsylvania movements have a high bell stand and often a large bell, figures 2.9 and 2.11.

Just at the turn of the eighteenth and nineteenth centuries there is evidence that some German-American clockmakers switched from casting and forging their own parts to making movements from "kits" imported from England. Almost overnight their massive movements with interesting features shifted from lantern pinions to cut pinions and conventional looking British layouts.

Thus one can find two clocks signed on the dial by a known German-American clockmaker from almost the same period with vastly different movements.

The brass movements from New England do not have the German heritage, but again they do have identifiable features. It is often today's local repairmen—who have seen many local clocks over a lifetime with unusual features—who are the best source of information about them. These features are often confined to a single maker or a family of makers. New England eight-day movements from the eighteenth century look much like their British counterparts. After 1800, movements from the northern New England states of New Hampshire and Vermont show much skeletonizing of their brass plates while some clocks from Maine, Quaker-made, have iron plates. These were efforts to reduce costs and attract business from a not-too-affluent public in these states.

In New Jersey and bordering New York the same was true: there was skeletonization and thinning of plates and other cast-brass parts, again to keep costs down. In New Jersey, barrels were sometimes made of wood covered with a thin sheet of corrugated brass which was nailed to the barrel. Cleverly, the seam was shifted one space in the corrugation making a spiral for the gut to follow on the barrel. In New York and New Jersey sometimes hammer-springs were coil springs made from a section of a broken watch spring. This is one modification that was a lower-cost improvement and should have been used more widely, figures 2.14, 2.15 and 2.16.

Returning to cases, there is an old saying about Pennsylvania grandfather clocks: "one tree, one clock." While clocks from outside Pennsylvania may not be as substantial, they are still stout by British standards. Boards are a full 1 inch thick. There is another saying that it is easy to tell an American from a British clock even when it's dark—just lift it. If you can't get it off the ground, it's American.

American grandfather clocks are heavy and are made of native American woods. They also differ from their British counterparts in form and construction. In the nineteenth century, American and Scottish clocks remained tall and thin. English and Welsh clocks became wide and less deep. American clocks, again like Scottish clocks, used a lot of dovetailing in their case construction. English and Welsh clocks relied upon glue blocks and nails. Pennsylvania clocks often had swan-neck pediments. New England clocks often had arch tops with frets.

In writing this chapter I have often compared American grandfather clocks with their British counterparts and, except for the German influence on Pennsylvania movements, I have not mentioned any other European country in regards to influencing American brass-movement grandfather clocks. The reason is—I couldn't find any such influences. When you look at an American grandfather clock it looks British and unlike any other floor-standing clock from Europe. The movements of most American grandfather clocks also look British. I have spent a lot of time pointing out the German-American characteristics of some movements, but they are the exception, not the rule.

Figure 2.16 (opposite page). New York State clock showing the German-American strike control with both the rack lifting piece and the rack hook pivoting from the right side of the front plate. The blobs of solder are poor repairs, the one in the center being to add weight and make sure the rack lifting piece drops.

CHAPTER 3
THE BANJO CLOCK

The Willard brothers grew up 250 years ago on a farm in rural Massachusetts about 35 miles west of Boston. Today the site is little changed and exists as the Willard House Museum of Grafton, Massachusetts (www.willardhouse.org) honoring the brothers' contributions to clockmaking.

It is all too easy to tell the story of the Willards with Simon Willard's ultimate achievement, the invention and patent in 1802 of the banjo clock, in a step-by-step manner as if it were a group of scientists developing radar in the 1930s, but this was probably not the case. There may have been underlying reasons why an almost untrained group of clockmaking brothers developed the first truly American clock, figure 3.1.

The Willards were a family of 12 children—nine boys and three girls—11 of whom reached adulthood. Four of the young men took trades common for the time: Joseph (b.1741) was a minister; Solomon (b.1745) was a tanner; John (b.1748) was a shoe last maker, and Joshua (b.1751) was a blacksmith. The other four sons became clockmakers.

The oldest of the four clockmakers was Benjamin (b.1743) followed by Simon, born 10 years later in 1753. Then came Ephraim in 1755 and last, the youngest son, Aaron, who was born in 1757. As an aside, the first and last two children were girls. Benjamin had left home by the age of 2l, working as a shoe last maker in clockmaker Benjamin Cheney's shop in East Hartford, Connecticut. Shoe lasts are the wooden frames that leather shoes are made on and, interestingly, Benjamin's five-year-younger brother, John, was also a shoe last maker. There is some evidence that the greater Willard family was involved in shoemaking. Clockmaker Benjamin Cheney made wooden-movement tall clocks so he probably had the tools that Benjamin Willard needed to make lasts and the connection of a clockmaker sharing a shop with a shoe last maker makes a bit more sense. Probably from working with a clockmaker, Benjamin Willard picked up some skills and knowledge of the trade.

It is an important point that Benjamin Willard, and in turn his three clockmaker brothers, Simon, Ephraim, and Aaron, did not serve proper apprenticeships as clockmakers and this may have encouraged them to think "outside the box" and create new and innovative products.

Benjamin returned home from Connecticut in 1766 and built a "clock shop" at the farm in Grafton, but he quickly departed to Lexington, Massachusetts, to work for Nathaniel Mulliken, a brass-movement clockmaker, allowing him to learn that side of the trade. That venture ended with the death of Mulliken in 1768, leaving 25-year-old Benjamin Willard still only lightly trained as a clockmaker. Simon (15), Ephraim (13), and Aaron (11) were back in Grafton working in the clock shop with almost no training.

Benjamin moved to several locations in nearby Massachusetts, including Roxbury on the edge of Boston. From these locations he would work for a short period advertising he had workmen, possibly he should have called them "workchildren," in Grafton.

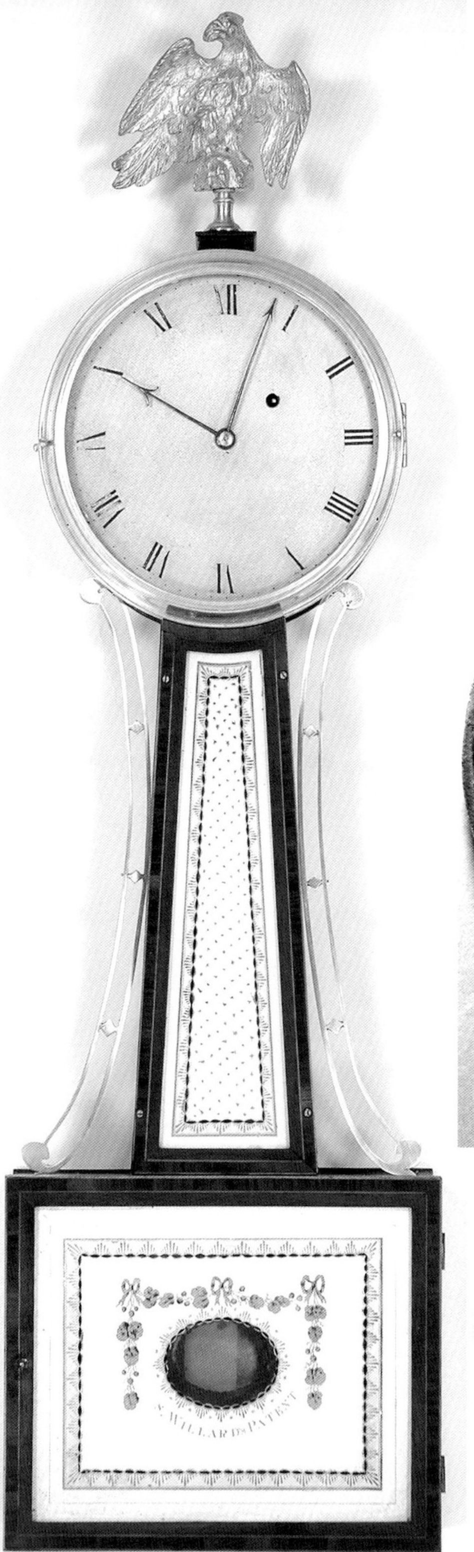

Figure 3.1 (left). Simon Willard, Roxbury, Massachusetts, mahogany "Patent Timepiece" or banjo clock with reverse painted glasses, ca. 1805-1810, height 34 inches. Photograph courtesy of Paul J. Foley.

Figure 3.2 (above). Lead pendulum bob from Willard tall clock, the bob cast with inscription "John Morris / S. Willard / Clock Pendulum / 1770 / 1771." Photograph courtesy of Paul J. Foley.

Figure 3.3 (above). Aaron Willard, Boston, Massachusetts, case-on-case shelf clock with painted dial, ca. 1800-1805. Photograph courtesy of Paul J. Foley.

Figure 3.4 (left). Simon Willard, Grafton, Massachusetts, mahogany 30-hour wall clock with brass dial, ca. 1777, height 24 inches. Photograph courtesy of Paul J. Foley.

Figure 3.5 (left). Simon Willard, Roxbury, Massachusetts, mahogany tall clock in "Roxbury-style" case, 12 inch Osborne / Birmingham painted dial, ca. 1795-1800. Photograph courtesy of Paul J. Foley.

Figure 3.6 (above). Eight-day brass movement from Simon Willard banjo clock in figure 3.1. Height 4½ inches. Photograph courtesy of Paul J. Foley.

Figure 3.7. "S. Willard's Patent" signature on reverse-painted glass banjo-clock tablet, ca. 1825. Photograph courtesy of Paul J. Foley.

One of these "workmen" was a strange individual, John Morris, a trained clockmaker. At this point, it must be emphasized that who John Morris was remains a bit of a mystery and maybe John liked it that way, as he was a shady character. One version is relatively straightforward. There was a John Morris who was born in Woodstock, Connecticut, on September 5, 1735, and he was an apprentice to James Batterson of Boston. A James Batterson came to Boston from England via Pennsylvania in 1707. He was trained by Robert Batterson, his father, in London in 1696.

We have to take a bit of a diversion here as the story of James Batterson is a bit too good to leave untold. Batterson presented himself to the town of Boston and their records of September 29, 1707, reveal "James Batterson, Clockmaker being presented Says he came from Pennsilvania [sic] into this Town abt a month Since & desires to dwell here, the Select men do now warn him to depart out of Town to finde Suretyes to save the Town from Charges." Batterson found "Suretyes" in John Smith and Thomas Thacher and twice advertised from Boston as "Lately from London" and " ... for new clocks or old ones turned into pendulums." That would mean he converted lantern clocks from balance wheel to pendulum, probably anchor, escapement. This James Batterson was the first Boston clockmaker I have documented. Unfortunately for our purposes, this James Batterson died in 1727 and could not have been the one that trained John Morris who was not born until 1735. James Batterson had a son, James Batterson II, born in 1718, who was also a clockmaker and he would have been 31 when he took the 14-year-old John Morris as an apprentice in 1749.

Simon Willard's obituary of September 9, 1848, (*Roxbury Advertiser*) sheds a bit of light and still more mystery on Morris. "He [Benjamin Willard] secured the services of a Englishman by the name of Morris, to impart a knowledge of the construction of brass clocks to himself and his brothers Simon and Aaron." So now Morris is an Englishman? We know Morris had a troubled youth and was in jail in Norwich, Connecticut, in the early 1760s. Later in 1777 clockmaker Elias Yeoman accused him of stealing silver and posted a reward for his apprehension: " ... the thief is one John Morris, a clock maker, and is well known in Hartford, Middletown and

Figure 3.8 (left). Elaborate gilt girandole banjo clock, ca. 1970, by Elmer 0. Stennes, Weymouth, Massachusetts. Height 44 inches. Photograph courtesy of Paul J. Foley.

Figure 3.9 (right). Sawin & Dyar, Boston, Massachusetts mahogany lyre-type banjo clock, ca. 1825. Height 41 inches. Photograph courtesy of Paul J. Foley.

Norwich [Connecticut]."

We have spent too long on John Morris but he and Benjamin Willard probably provided all the training Simon, Ephraim and Aaron would receive. Morris did leave behind a group of cast lead pendulum bobs dated "1770-1771" and signed by John Morris and Simon Willard, figure 3.2. These bobs are found on a few early brass-dial Benjamin Willard tall clocks.

From here on, the story of the development of the banjo clock is the story of Simon Willard. In 1771 Simon, then 18, was working at the Grafton clock shop with his brothers Ephraim and Aaron and with Mr. Morris. Benjamin had his shop in Roxbury near Boston. In 1775 Simon and at least his brothers Ephraim and Aaron, and probably some of his other brothers, marched to Roxbury for military service in the Revolutionary War, but they seemed to be quickly back at Grafton, making clocks, tall clocks. By 1777 Simon was making the "Grafton 30-hour brass-dial wall clock," figure 3.4. This was the first step towards the banjo clock.

His next step was not directly towards the banjo clock, but it resulted in a most beautiful American clock, the Massachusetts shelf clock. In Chapter 1, I explained that one of the problems for American clockmakers was that Americans could not make springs. The Massachusetts shelf clock is a prime example of making what looks like a British spring-driven bracket clock but, by building an attractive box under the clock for them to fall through, it could run on weights, not springs. Though not solely the invention of Simon Willard, he was among the very first to make clocks like this. It was an eight-day clock and the first to have the winding square at two o'clock, figure 3.3.

By possibly as early as 1780, but for sure by 1783, Simon and Aaron were working in Roxbury on the very edge of Boston. In 1792 Aaron moved a few blocks, just into Boston. Aaron, it would turn out, was a great businessman and gathered great craftsmen: not just clockmakers but cabinetmakers; dial and reverse glass painters; and others, to work at his manufactory. Simon, while also a good businessman, was a great innovator.

Simon received his first patent in July 1784 for a "clock jack" (roasting jack) and a very interesting group of letters exists between Simon Willard and the American patriot Paul Revere where Willard is pleading to Revere for money and asking him to take more of his roasting jacks.

Simon and Aaron Willard were among the first clockmakers in America to control all aspects of the production of their clocks. They would control the manufacture of movements, cases, dials, hands, weights, etc., and present the clock as a finished product for sale at their shops. This practice probably grew out of the manufacture of their 30-hour wall clocks and Massachusetts shelf clocks, where all the components, even weights, had to work in harmony. The practice probably spilled over to the tall clocks they produced. Later in their careers they even purchased complete tall clock movements from England, but in my opinion, not on a large scale. They still took orders for clocks and bespoke items, but many customers just bought "off the floor." This led to a great similarity of the clocks they sold. At a glance, Simon's and Aaron's tall clocks looked very much alike. Cabinetmakers in Roxbury and Boston made cases for them, but they all looked very similar, and today the case is referred to as the "Roxbury" case, figure 3.5.

The town of Boston, by then possibly a city, existed on a peninsula with only a small neck of land connecting it to the rest of Massachusetts and New England. On that narrow neck was one road with a few shops at a wide spot and that spot was Roxbury. Travelers to Boston from anywhere else by land would have to approach Boston over this narrow neck and along the way, with Boston in sight, pass by the wonderful clock shops of Simon and Aaron Willard. Many of these visitors had traveled a day or two and if they stopped and purchased a clock they could take it away with them.

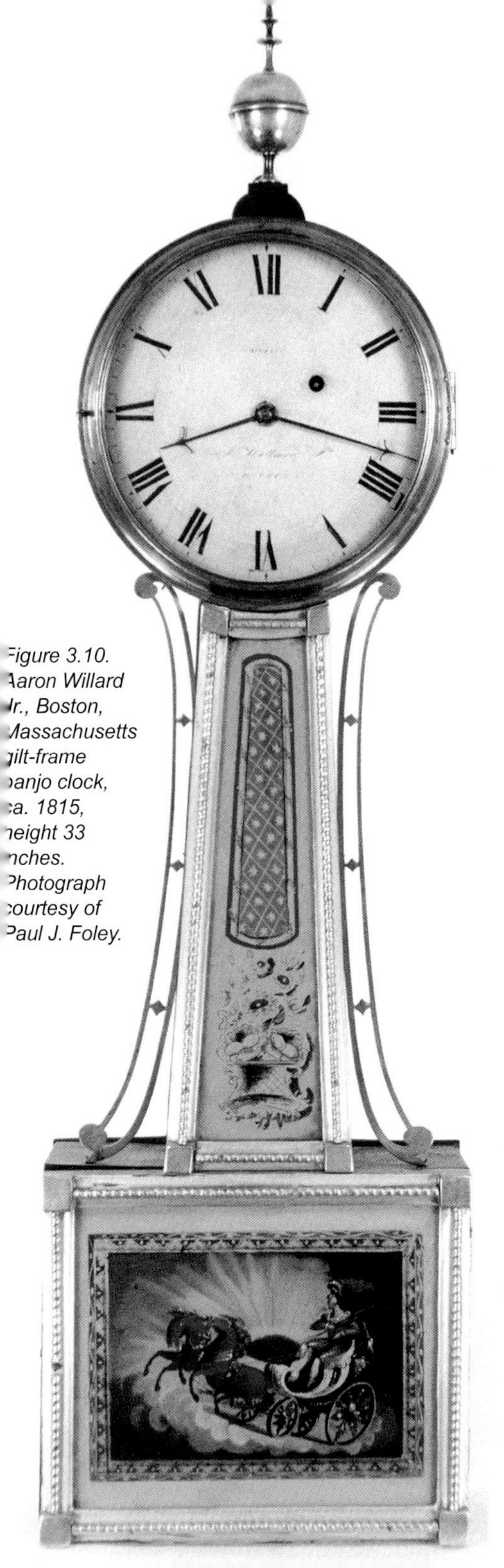

Figure 3.10. Aaron Willard Jr., Boston, Massachusetts gilt-frame banjo clock, ca. 1815, height 33 inches. Photograph courtesy of Paul J. Foley.

By 1800 things were going well for the inventive Simon Willard and he was beginning the development of what was to be his "Improved Timepiece." A timepiece is a device that shows the time but does not strike. A clock, technically, is a device that shows the time and sounds the hours. The "Improved Timepiece" is today's banjo clock.

The heart of a banjo clock is a time-only eight-day movement about two-thirds the size of a tall clock movement with a three-quarter second pendulum. The suspension has been moved from the back to the front of the movement and all is driven by a narrow weight shaped to conform to the taper of the case. The great wheel is wound at two o'clock to compress the train layout and allow the great wheel to clear the bottom plate pillar. None of this is seen, figure 3.6.

Constructed around the functional movement, weight and pendulum is a case, the form of which follows the functional parts as closely as possible. The secret of the banjo's success is the decoration of the parts that are seen: glasses, inlay, brass side-arms, hands, bezel and finial. The original beauty and present condition of these parts significantly affects the price of the clock on today's market, by at least a factor of 10 or 20 times, and maybe even 100. The wonder of the banjo clock with all its attention to detail is that it has never gone out of fashion with probably more weight-driven banjo clocks being made in the last quarter of the twentieth century than the first quarter of the nineteenth. The methods used to make these modern banjo clocks are essentially the same as they were 200 years ago.

Successful with obtaining a patent for his roasting jack, Simon Willard applied for and received a US patent for his banjo clock signed by then president Thomas Jefferson on February 2, 1802.

Simon Willard knew the value of any patent item and proudly displayed "Patent" in association with his new clock whenever he could, so much so that "Willard's Patent Timepiece" became one of the names used to describe the clock. "Patent" was often painted into the glass designs, sometimes as "S. Willard's Patent," figure 3.7. The term "banjo clock" is in almost universal use today. Its use became popular about 100 years ago.

Production of the banjo clock was centered around Roxbury and Boston with probably half or more of the early clocks being made there. Other centers were Concord, Massachusetts, maybe producing 20 percent or so and Burlington, Vermont, a little under 10 percent. The rest were made here and there in New England.

There were two major variations of the early banjo clock: by early I am referring to the first quarter of the nineteenth century. The variations were the girandole and the lyre clock.

The most impressive form of the banjo clock was the beautiful girandole, figure 3.8. This was an invention of Lemuel Curtis who received a patent on July 12, 1816, for an "Improvement upon Willard's time-piece." It became known in the period as the "Curtis Patent Timepiece" and again this name didn't stay with the clock. Girandole mirrors were popular at the time and the round form of the base of the clock, with its convex glass surrounded with gold balls, looked much like the girandole mirror, so the name girandole became applied to the clock.

The last banjo variation was the lyre timepiece, a product of (John) Sawin & (George) Dyar of Boston, and was first seen in about 1825, figure 3.9. An advertisement by Sawin in 1833 describes them as "Elegant Harp Pattern Time Pieces," but that name also didn't stick and they are lyre clocks today.

From this point, 1830, less than half the story of the banjo clock has been told. The clock continued to change throughout the rest of the nineteenth century and into the first half of the twentieth century. America's railroads were regulated by later forms of banjo clocks still being made after World War 2. Also omitted in this chapter is the fact that Simon Willard made gallery (dial) clocks, tower (turret) clocks, regulators and a curious lighthouse clock for which he received a patent in 1816/1817.

If you are interested in purchasing an early banjo clock—beware! Probably second only to English lantern clocks are banjo clocks in terms of being difficult to understand to the level required to purchase one on the open market, such as an auction.

If you want a good banjo clock you are going to have to spend a great deal of money. The difference between a great $100,000 banjo and a good one that costs $5,000 is hardly apparent to the novice. Never buy an early banjo clock on the Internet, though the Internet is a good place to go to view many such clocks and see trends in prices.

A sensible alternative is to buy a high-quality new banjo clock. These clocks are reasonably priced, run well and are wonderful to look at. Their value holds up. One maker's clocks of the 1950s and 1960s are about the same price as an average clock from the 1820s. It's a strange world.

CHAPTER 4
THE WOOD MOVEMENT TALL CLOCK

Why make a clock movement out of wood? Why not brass or iron? These are often-asked questions concerning America's wooden-movement clocks. For years the answer given was that brass was in short supply in the young country and clockmakers were forced to use wood. The truth is that brass was simply expensive. It always has been. It is today. Wood, on the other hand, is cheap.

Brian Loomes, an acknowledged expert on British clocks, writes about early English 18th century clockmakers, often rural clockmakers, using iron in place of brass in their clocks. That was done for much the same reason that wood was used in America. It kept the cost down and allowed the maker to reach a different market, a lower income market that couldn't afford a brass clock. As an aside, Mr. Loomes has also told me that he has come across records of a few 17th century British wooden clocks in pre-1660 house inventories, but he has never seen such a clock.

Besides wood being low in cost, it has three advantages over brass or iron in making early mass-produced clocks. The first is that wood is self lubricating. Goose grease or pig lard both break down, smell bad, and quickly lose their lubricating properties. That is why Englishman James Harrison used wood to make his famous wooden regulators in the 1720s.

Secondly, wood machines easily. The

Figure 4.1. Eli Terry "Porter Contract" movement No. 3406. These movements are numbered on the back of the dial and on the seatboard, as this one is. There were 4,000 movements made; this one would have been made in the final year, 1809. Photograph courtesy Carter Harris, Headquarters National Association of Watch Collectors, Columbia, Pennsylvania.

Figures 4.2a to 4.2c (this page and facing page). Thomas & Hoadley wooden-movement tall clock. Eli Terry sold his factory to his two journeymen Seth Thomas and Silas Hoadley who were in business together from 1810 to 1813. Photographs courtesy Cowan's Auctions, Inc., Cincinnati, Ohio.

earliest makers, such as Eli Terry and his competitor James Harrison (that would be the American James Harrison from Waterbury, Connecticut and not the one from England mentioned above) knew that high speed cutters—that's high speed by 1805 rural-American standards—could be made to machine wood but probably not brass and certainly not iron.

The last factor was that wooden clocks were already a proven commodity. Hand-made versions had been known in America from before 1750. Two or three generations had grown up around them and there was no fear in a New England farmer when he was thinking of purchasing a wooden clock. He simply didn't have the money to buy a hand-made one.

As we saw in Chapter 1, Eli Terry of Plymouth, Connecticut, trained as both a brass and wooden clockmaker, had been successfully making wooden movements in batches of about 25. Within those batches he had achieved interchangeability of parts. In nearby Waterbury, Connecticut, James Harrison had been doing the same thing at the same time, 1805, in about equal size batches, but he had not achieved interchangeability.

Terry received financial backing from two unrelated merchants named Porter in 1806 to produce 4,000 wooden tall-clock movements for $4 each. Terry sold his existing 20 foot square factory to his former employee Heman Clark and purchased land still along the Naugatuck River and built the "Ireland" factory. He hired two 21-year-old joiners, Seth Thomas and Silas

Hoadley, tooled the factory, and set about working on the "Porter contract" to build the 4,000 movements. He completed the contract in 1809, figure 4.1.

The "movements" Terry produced included the dial and hands, a seatboard that was an integral part, weights and a pendulum—everything, in fact, but the case. The tin weight shells were shipped empty to be filled by the purchaser. From now on in this chapter the word "movement" will be used to refer to all of the above. At the time, all these parts, minus the case, were referred to collectively as a "clock." The whole thing with the case was also, confusingly, referred to as a "clock."

So how much did a brass 30-hour tall-clock movement cost? Working backwards, the whole eight-day clock cost $65, almost an industry standard that lasted for years. Simon Willard sold his top-of-the-range tall clocks for $100. A 30-hour brass clock would have cost less; it had a lower-quality case and a lower-cost painted dial. Guessing, using the same ratio of costs as in England (which we already know about) and the $65 cost as a basis for an eight-day clock, a 30-hour brass clock probably cost about half the price of an eight-day clock, about $33. Breaking that down into movement, case and dial—and including a rather large guess factor—it might

Figures 4.3a and 4.3b (facing page). Riley Whiting clock. The case is probably poplar or pine and retains its original finish. Most readers outside America would be shocked to learn that this clock with nothing going for it except its original condition was the third-highest-priced clock in a sale of 400 clocks. All the clock collectors were shocked, but antique dealers knew what their customers wanted. Photographs courtesy Cowan's Auctions, Inc., Cincinnati, Ohio.

have been about $18 for the movement, less than $10 for a simple case and about $10 for the dial. Remember that Terry's $4 movement included everything but the case. If that price is about right, the wooden movements Terry could sell for $4 were cheap indeed.

After the above calculation was made, a period source was found that confirms the numbers. From *Eli Terry and the Connecticut Shelf Clock* by Roberts and Taylor, page 46, Eli Terry's son, Henry, states in 1853, looking back at the prices in the 1790s, "The price of brass clocks was from £10 to £15 or $33.33 to $50. This was the price without a case. The case might be procured at a price varying from $5.00 to $30.00 according to the quality and materials of which it was made: so that the entire cost of a [hand-made] wooden clock with the case, was from $18.00 to $48.00 and for brass clocks $38 to $80." Assuming the $38 was the cost for a complete 30-hour brass-movement clock, that compares well with our estimate of $33.

Why stress the cost of the mass produced wooden-movement clock that Eli Terry made? The answer is the thrust of this chapter. The clock was so inexpensive to manufacture—note the use of the word "manufacture" and not "make"—that it opened a whole new market. The rest of the world made clocks for the well-to-do; Americans manufactured clocks for anyone who worked. In the early 1840s, newer models of these inexpensive, mass-produced clocks from America, now with brass movements, would spread across the world, sadly ending production of the hand-made clock.

Terry's manufactured clocks were on the market by 1810 and were an immediate success. On the surface one would expect that he would revel in that success, sit back and enjoy life, and let the profits roll in. Not so. Almost immediately, in 1810, Terry sold his factory, and its tooling and manufacturing rights, to his journeymen Seth Thomas and Silas Hoadley. It was as if Henry Ford sold his car factory just as the first Ford came off the assembly line.

Why?

Terry realized there was a fatal flaw in manufacturing his mass-produced tall clock movements. The movements left the factory and were distributed to the customer without a case. At the point of sale, often a farm, the peddler had to convince the farmer that it wouldn't be hard for him to go to town with the movement and find some cabinetmaker to make him a case. It was as if Henry Ford had produced his car without a body. Terry also realized that the tall clock was the wrong clock to mass produce and sell in America with its almost unlimited frontier and no system of roads. The country was too big. The clock was too big. A much smaller clock with a case was the answer.

Terry had the idea of producing a cased shelf clock ready to run when it left the factory. No longer would local cabinetmakers take a share of the profit. It would all go to Terry—and the merchants and peddlers who sold the clocks.

In 1810 Thomas & Hoadley were manufacturing wooden tall clock movements in Plymouth, Connecticut, figures 4.2a to 4.2c. In 1806 James Harrison of Waterbury, Connecticut, was duplicating the work of Eli Terry in nearby Plymouth. Harrison had installed water power to his factory and was making wooden movements in batches, but not with interchangeable parts, when in 1806 he sold out to William Leavenworth who went on to quickly make mass-produced wooden-movement clocks at Waterbury, but not as successfully as Thomas & Hoadley. William Leavenworth relocated to Albany, New York, in 1815. Mark Leavenworth, his nephew, built his own factory in Waterbury and successfully manufactured wooden movements. Lemuel

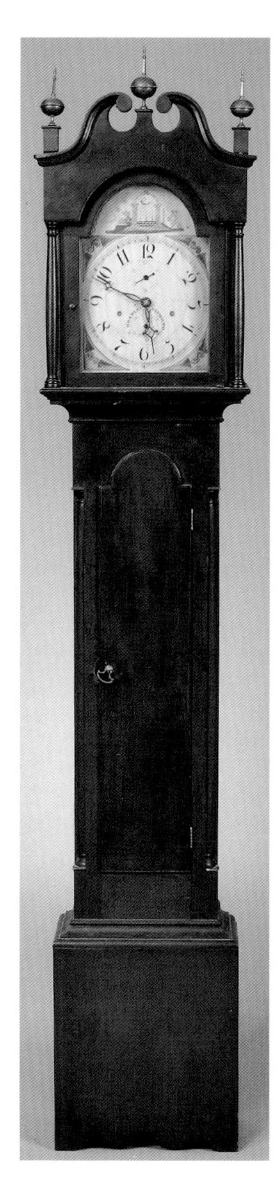

Harrison, James's brother, had a successful factory in Waterbury in 1811.

Clockmakers Samuel Hoadley and Luther Hoadley, both distant relatives of Silas Hoadley, along with brother-in-law Riley Whiting, had moved from Waterbury and were manufacturing wooden movements in Winchester, Connecticut, 40 miles from Waterbury, starting in 1807 as Hoadley's & Co. Luther died in 1813 and the firm became Samuel Hoadley & Co. Riley Whiting became owner in 1819 and it was Riley Whiting who turned out, along with Silas Hoadley of Plymouth, to be the major Connecticut manufacturers of wooden tall clock movements who lasted into the 1830s, figures 4.3a and 4.3b and 4.4a to 4.4c.

Figures 4.4a to 4.4c (this page and facing page). Silas Hoadley clock with Masonic symbols in arch of dial. Thomas & Hoadley dissolved their partnership in 1813 and each man began manufacturing on his own. Photographs courtesy Cowan's Auctions, Inc., Cincinnati, Ohio.

The first clockmaker in Bristol, Connecticut, was Gideon Roberts and by 1810 he, along with his sons Elias and Gideon Jr., each had shops in Bristol. The three men made tall clock movements in considerable numbers but none had water power in their shops. Gideon Roberts and his son Gideon Jr. both died in 1813. The Roberts are important as the first clockmakers in Bristol, then a town of about 1,300 people, the place that was to become the clockmaking center of the world only 30 years after Gideon's death.

The manufacturing of tall clock movements quickly spread from Connecticut. Two centers sprang up 700 miles west in Ohio, figures 4.5a and 4.5b. Montreal, Canada, was a major

Figures 4.5a and 4.5b. Read & Watson wooden-movement tall clock ca. 1814. Note this partnership has no town name on the dial. Between 1811 and 1814 Read & Watson were peddling Connecticut clocks in Ohio with their names on the dial. They were having cases made in Ohio if the customer could afford a case. In 1815 Read & Watson located in Cincinnati and built a pony-powered factory to make their own movements; later the firm became just Luman Watson. Photographs courtesy Cowan Auctions Inc., Cincinnati, Ohio.

Figures 4.6a to 4.6 (opposite page, Orange Hopkins cloc movement from Litchfielc Connecticut ca. 182(The hands are pewte painted black. The secon hand and calendar han are missing. The windin holes are only painted o the dial. The bell shoul sit above the movemer and has been removed fo these photographs

center making cases and sometimes assembling movements, but most of the time they used Connecticut-made movements for their Canadian cases.

The manufactured wooden-movement tall clock had a 25-year span of popularity ending in about 1835, just at the start of the first great American depression. The popularity of the clock reached its peak in about 1820 and then went into decline after Terry's shelf clock, the pillar-&-scroll, appeared on the market. Wholesale prices of the wooden tall clock movements started at $4 and then declined along with the popularity of the clock until, by about 1833, prices were approaching $1. Almost all of the clock companies producing shelf clocks and tall clocks went under during the depression of 1836/1837 and only a handful came to life afterwards. None who returned manufactured tall clock movements. There was one partial exception. Riley Whiting in Winchester died in 1835 and it seems his wife and son continued some sort of minimal operation, possibly producing movements or parts for the Canadians.

The wooden movements used specific woods for each of its parts. Plates were made of well-seasoned, quarter-sawn red oak. Wheels were made of quarter-sawn American cherry, again

well seasoned. Pinions and the pinion arbors were made of New England laurel, a small, slow-growing—now endangered—plant that grew along the ground. Pieces varied in diameter from thumb- to wrist-size. Pivots were wire inserted into the ends of the laurel pinion arbors. The fly was lightweight poplar and the seatboard was usually pine. Dials were poplar. The escape wheel was brass and the anchor, or verge, was steel.

The clocks often had the maker's name on the dial, but many were unsigned. Other than Canadian clocks, the cases were made by cabinetmakers near the point of sale. The New England cases were often pine, of simple construction, and almost always painted when new. Ohio clocks were much more sophisticated, usually made of cherry. Some simple Ohio cases were painted poplar. Montreal cases were factory-made with identical case parts made of pine and painted, differing only in color. They sometimes had a label in French. Very few other cases had labels.

For today's collector, the paint-decorated clocks are very expensive, up to $50,000 for great examples. Good hardwood clocks are from $2,000 to $10,000. The run-of-the-mill pine and

Figures 4.7a to 4.7e (this page and facing page). Wooden movement tall clock, unsigned, as many were. The case is probably from Pennsylvania and shows the enthusiasm of the owner and cabinetmaker. Photograph courtesy George and Cathy Goolsby, Houston, Texas.

poplar clocks with stripped cases are from $800 to $2,000, and cut-down examples sell for the price of the movement.

The most common cause of failure is a tooth snapping off when the clock is forced to run on heavy weights. The clocks should run on two $3^{2}/_{3}$ pound weights, the same amount of weight as used to power an OG.

We will conclude with a testament to the quality of these wooden-movement tall clocks. About 25 years ago I purchased a wooden-movement clock that was in a barn, rotted into the muck. Fortunately, the movement was 5 feet off the ground, in place in the clock. Within a week I sold the clock cheaply to a lady with the understanding that "it will probably never run." Later that day the lady called back and said it was up and running. I see her about once a year and with no servicing—remember these clocks are self-lubricating—the clock has run for the past 25 years. But that is an exception.

Terry's mass-produced wooden tall clock movements represent one of America's greatest contributions to horology: the manufacturing of clock movements.

CHAPTER 5
THE PILLAR-&-SCROLL

The year was 1810. The young country of America was expanding westward at an unbelievable rate, an uncontrollable rate. In 1776, when America declared independence, half the population lived within 13 miles of saltwater—that does include rivers that were brackish and affected by the tides, and there were many such rivers. The rugged forests and mountains seemed impenetrable.

By 1800 Americans understood that the small westward-flowing streams they discovered deep into the mountains would take them to vast rivers in the West. Settlers were pouring west across the mountains on rugged trails into the new territories, and France had just sold America a vast tract of land that she had only recently procured from Spain. That land extended all the way to Montana. Tracts as large as 400 acres of the world's best farmland could be claimed by the first man there. Often they were the very young, sometimes lads of 21 who just packed up and went west. In a year or two of very hard work they had made it. They were on the way to being wealthy. By the tens of thousands they were ready … to buy a clock.

Back in Connecticut, 38-year-old Eli Terry had successfully completed his "Porter Contract" and then sold his old factory, called the "Ireland" factory, to Seth Thomas and Silas Hoadley for $6,000. Terry now set out to develop and produce his new shelf clock, a clock that would leave the factory complete with a case, a clock he could sell across America.

Terry had already received some of the $16,000 he earned from the Porters for producing the 4,000 wooden tall clock movements upfront to set up his Ireland factory. The final payment from the Porters was received in land, and it took Terry a couple of years to liquidate those properties for $7,000. This meant that Terry had $13,000 of capital available for his shelf clock project.

He purchased a seven-acre mill site on the same Naugatuck River in Plymouth, Connecticut, this time near the Reynold's Bridge, on December 14, 1812. In 1813 he added several more acres of nearby land and in early 1814 purchased the rights to build a 14 foot dam to power his factories. In March of 1814 he purchased a nearby home for his family.

During this period he began to experiment with what today we would call prototypes of his new shelf clock. His concept, which proved successful, was to produce for the first time a completely mass produced clock, both case and movement, which could be sold to peddlers or merchants at the factory, ready to run.

The ultimate production clock was a pillar-&-scroll case with a five-arbor movement, which was on the market in 1821 or 1822 and was patented on May 26, 1823. The production case was about the size—with some of the design elements—of a grandfather clock's hood. It had a delicate swan-neck pediment, referred to as a scroll top, with three usually brass finials on three plinth blocks. The mahogany veneer on the scroll top ran vertically while the poplar secondary

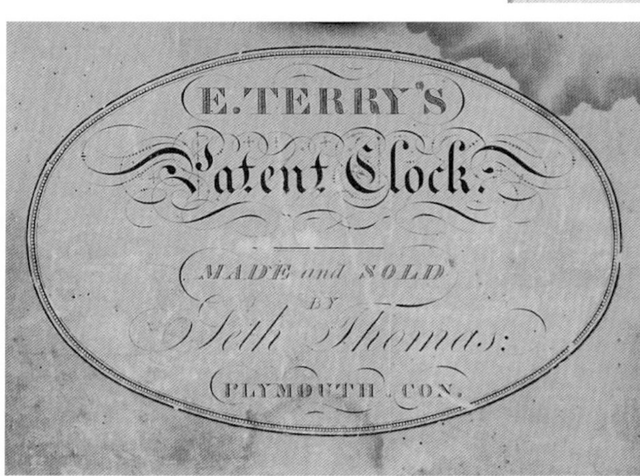

Figure 5.1a (above left). Experimental pillar-&-scroll box clock, ca. 1816. The case is the size of early pillar-and-scroll clocks with the scroll top, side pillars and bottom removed. Note the front piece of glass has no division between the top and bottom sections and the clock's numerals are painted on the back of the glass (reverse painted). With permission of the clock's owner, the American Clock & Watch Museum, Bristol, Connecticut.

Figure 5.1b (top right). Model 1C wood-strap front plate, rack-and-snail movement that uses the center board of the three backboards as the back plate of the movement. Note the piece of angled brass that extends upward from the bottom wooden strap of the front plate of the movement; it supports the center wheel and gives a pivot point for the rack. Of the 14 known Model 1 clocks, 10 are Model 1C. This is the movement that Terry patented in 1816.

Figure 5.1c (above). Seth Thomas and Eli Terry worked together on the development of the pillar-&-scroll. Their factories were very close together. This experimental clock carries Seth Thomas's rather than Eli Terry's label.

Paten[illegible]cks,

Made and Sold by

Seth Thomas;

And warranted if well used.

DIRECTIONS

For setting up and Regulating said Clocks.

Let the clock be set in a perpendicular position: this is necessary, in order to its having an equal beat: and if it fails to beat equally, it may be put in beat by bending the wire which receives the pendulum. If the clock goes too fast, lengthen the pendulum; but if too slow, shorten it by means of a screw at the bottom. If the hands want moving, do it by means of the longest; turning at any time forward, but never backward; and if the clock fails to strike the proper hour, pull downward the small wire which hangs down under the face on the left side of the clock, and it will strike as often as you repeat the motion. Apply no oil at any time to it, unless it be a very small quantity of the oil of almonds, or sweet oil to the points of the teeth on the brass, or crown wheel.

To wind up the weights, put the key on with the handle down, and turn it towards the figure 6.

N. B. The public may be [illegible] at the above factory, are equal, if not superior, to any made in this country.

Pl[illegible]th, Conn[illegible]

Figure 5.2a (above left). Production Model 2, the so-called off-center pillar-&-scroll, ca. 1819. Note the bottom glass, which is now separated from the top glass with a wooden divider, has an opening to show the pendulum. The pendulum is to the right of the center-line of the clock or "off-center." With permission of the clock's owner, the National Clock and Watch Museum, Columbia, Pennsylvania.

Figure 5.2b (top right). The Model 2 movement has countwheel rather than rack-and-snail strike control. Note that the movement has no solid center wheel support from the front plate. The large countwheel seems to be hanging out in the open, with only a wire to connect it to the front plate and support it. This was a fatal flaw of the Model 2.

Figure 5.2c (above). Seth Thomas label. Thomas contracted with Eli Terry to produce this clock with a royalty fee of 50 cents per clock. Terry would later claim that this contract did not apply to the later Model 5 clock and that Thomas had never paid him the 50 cents per clock fee for the Model 2 clocks Thomas produced.

wood, probably only air-dried, ran side to side; this caused the swan necks to curve backward. If the swan necks do not bend backward, the fragile top has been replaced.

Two fine columns or pillars ran the height from top to bottom. Early production clocks had rectangular wooden dials about 11 inches by $12^{1}/_{2}$ inches wide, while the final clocks had square wooden dials about 11 inches by 11 inches and beneath the dial was a reverse-painted glass or "tablet," recently made popular by the banjo clock. The feet were of a Hepplewhite design and the entire base was attractively scrolled. The combination of the side pillars and the scroll top gave rise to the clock's name.

The whole front surface of the clock showed quality mahogany veneers, sometimes with accents of maple veneers on the three plinth blocks at the top. The secondary wood was poplar. Internally the clock had horizontal backboards, usually three, with two internal rails or partitions about $^{3}/_{8}$ inches thick and 3 inches deep. These rails were mortised into the top and bottom. Their purpose was fourfold. They stiffened the whole clock. They provided a framework between which to mount the movement. They provided two channels outside the rails for the weights to run down, and with the addition of a stick across the rails at about the middle, they provided a means to secure the dial. On the top of the clock were two pulleys, one on each side for the cotton, possibly linen, weight cords to run through.

Terry determined that the clock's movement was to be made of wood like his grandfather clock movements. The seatboard of the grandfather clock movement was dispensed with as the case rails held the movement. The seatboard in the grandfather clock also had a secondary function of separating and securing the bottom of the front and back plates. Three bottom plate pillars were added to the pillar-&-scroll's movement to perform that function. The clock was to be weight driven as the Americans could not make springs. The two cast-iron weights, round in cross section, each weighed about $3^{3}/_{4}$ pounds, the same amount of weight as the grandfather clock, and because their weight cords ran up over pulleys at the top of the case, their force on the movement was upward. The weights could fall the full height of the inside of the case, about 21 inches, minus the 4 inch height of the weight. The clock had a duration of 30 hours on 17 inches of drop—or just over $^{1}/_{2}$ inch of drop per hour.

Terry's grandfather clock movement weights dropped about 5 feet or 60 inches in 30 hours, or 2in per hour. Compounding the problem, his earliest shelf clocks had a half-second and then the final clock had a .476-second pendulum while the grandfather clocks beat seconds. Without modifications to the movement, this increase in pendulum rate alone would also have required double the weight drop. Terry's new movement design accommodated the short drop.

Terry's grandfather clock and his shelf clock both suspended the weights on cotton cords. The grandfather clock was rewound by opening the trunk door and pulling down on the counterweight cord, thus raising one of the two weights. The shelf clock was wound with a key through the dial. The dial of the grandfather clock was often painted with faux winding holes—they are not real.

The doyen of the American shelf clock is Dr. Snowden Taylor and I would like to thank him for his help in preparing this chapter. His major work on the subject, along with the late Ken Roberts, is *Eli Terry and the Connecticut Shelf Clock*, 2nd edition. I mention Dr. Taylor's book because I cannot devote nearly the space and detail in this chapter to the development of Terry's shelf clock movement that would be required for the reader to come away with anything but a cursory understanding of the movement.

During the very early development of the pillar-&-scroll, Terry used a rectangular "box" version of his shelf-clock case, figures 5.1a-5.1c. These cases had the dial numerals painted in the upper portion of the door glass, probably to allow the observer to see the clock's movement in operation, and to allow easy access to the movement. The door glass was in one piece while

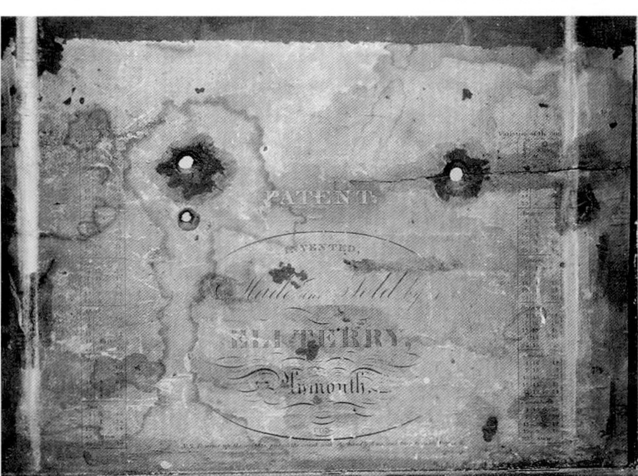

Figure 5.3a (above left). Production Model 3, outside-escapement pillar-&-scroll clock, ca. 1818. This was the clock Eli Terry was producing at the same time Seth Thomas was producing the Model 2. The dial of this clock was a part of the movement. The escapement showed while viewing the clock. The rest of the movement was attached to the back of the dial and not directly to the case. With permission of the clock's owners Cathy and George Goolsby.

Figure 5.3b (top right). Model 3 plated movement as shown from the back. In the background, the dial has two cleats attached to it on the left and right. The movement is mounted between these dial cleats. The bell, below, is also mounted to the dial.

Figure 5.3c (above). Worn Eli Terry label in the outside-escapement pillar-&-scroll clock.

the final production clock had the front door divided into two glass elements separated by a strip of wood. A few of these box-cased clocks, now with regular dials, found their way into later production, possibly as a cheaper version of the clock. All box-cased clocks are now very very rare.

By 1815 Terry was deep into the development of his pillar-&-scroll movement. Very likely some or many of his early attempts, the failures, have not survived. What have survived are 14 movements of a type known as the Model 1. These are further subdivided into four sub-categories: we will not delve deeply into their minor differences. All of these clocks have rack-and-snail strike control and a strap front plate—four thin strips of wood mortised together to form a sort of vertical rectangular picture frame. The frame having no center, the center arbor is supported at the front in all but two examples by a single brass piece support that sticks up from below and is retained by a "question mark" wire. The single brass piece support was a very bad arrangement and was the fatal flaw of the strap-front-plate movement, although it survived to go into early production.

The Model 1A clock was probably a patent model and need not be discussed. The 1B movement used the entire one-piece backboard of the case as the backboard of the movement. Pivot holes were drilled in the backboard. The 1C movement, representing 10 of the 14 known movements, had a three-piece horizontal case backboard, and the centerpiece of the backboard, extending from side to side of the case, was the back plate of the movement and was dovetailed into the back of the clock. The Model 1 was the movement Terry patented on June 12, 1816. Terry would claim that this patent applied to other future models he came up with, in particular the Model 2.

Still in 1816, Terry produced his Model 2 movement. This movement changed the strike control from rack-and-snail to countwheel control. There was an overlap in time between the Model 2 and the Model 1, as if development was going on with both models at the same time. The Model 2 clock used the back plate of the movement, which ran from side to side of the case, as the middle one of the three horizontal backboards of the case, like the Model 1C but with no dovetailing. The Model 1 and 2 had the pendulum suspended to the right of the center line of the movement. The clocks often showed the swinging pendulum through an oval opening in the bottom glass reverse-painted tablet of the clock and it was evident to the observer that the pendulum was not swinging in the center of the clock. The production Model 2 clock is today known as the off-center pillar-&-scroll, figures 5.2a-5.2c.

A sub-category of the Model 2 movement, the Model 2A is found in pillar-&-scroll clocks that had a hole cut into the face of the clock over the escape wheel so the motion of the escape wheel could be seen.

Seth Thomas, Eli Terry's former journeyman, was well aware of Terry's work and contracted with Terry on October 2, 1818, to produce Terry's Model 2 movements and sell them in his, Seth Thomas's, cases. Thomas was to pay Terry 50 cents for each movement he made after the first 800. It would seem Thomas had been making and selling clocks with Terry's blessing for years, very possibly in exchange for other work performed. The first 800 probably cleaned the slate for both individuals.

There were two more four-arbor movements, the Model 3 and the Model 4, overlapping with the Model 2 in particular. Both of them had solid front plates rather than the open-strap front plate found on Model 1 and Model 2 movements. The Model 3 had the escape wheel on the face, figures 5.3a-5.3c. The Model 4 went back to the escape wheel on the movement front plate. The version with the escapement on the face, the Model 3, became known as the "outside escapement," and the version with the escape wheel on the movement, the Model 4, became known as the "inside outside escapement."

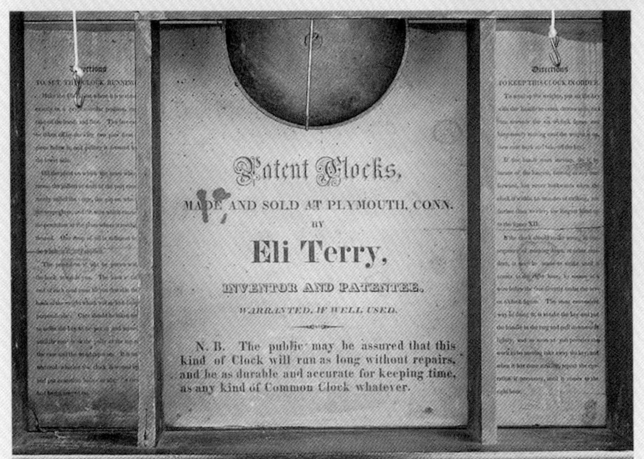

Figure 5.4a. The final production pillar-&-scroll, the five-arbor Model 5, ca. 1822. With permission of the clock's owner, the American Clock & Watch Museum, Bristol, Connecticut.

Figure 5.4b. While still a 30-hour weight-driven movement, the Model 5 had five arbors rather than four. It was the first model that had the movement mounted between two stiffeners that ran from the top to the bottom of the case. This was Terry's very successful model that he patented, but could never stop other Connecticut firms from making. Sometimes others used clever modifications to avoid the patent.

Figure 5.4c. Like-new label in the Eli Terry clock. Note the "Inventor & Patentee" on the label.

For a few years after 1818 Seth Thomas was manufacturing Terry's Model 2, off-center pillar-&-scroll and Terry was manufacturing his Model 3, outside-escapement, pillar-&-scroll. Judging by the number of extant clocks of each model, Thomas out-produced Terry. Terry was no doubt continuing to develop his five-arbor Model 5 and that was consuming much of his time. These two clocks, the Model 2 and Model 3, are rare and probably the earliest pillar-&-scroll clocks even a very serious collector will ever obtain.

Terry renegotiated his 1818 contract with Seth Thomas in 1822 and for $1,000 agreed to let Thomas make and sell the old Model 2 movement. Terry claimed Thomas had never paid him any of the 50 cents per movement royalty he promised in the 1818 contract. Thomas later went ahead and started manufacturing the new Model 5 movement anyway, causing the "Terry-Thomas" patent suits.

The Model 5 five-arbor-train movement was the culmination of Terry's work, figures 5.4a-5.4c. It was introduced in 1821 or 1822 and patented on May 26, 1823. The five-arbor-train became the industry standard for wooden shelf-clock movements for the next 20 years. This was the first clock to have the two rails or partitions inside the case. The case for the five-arbor movement would be 1.5 inches taller than the previous models and used two pulleys at the top.

The introduction to this chapter claimed the pillar-&-scroll was significant as the world's first completely mass-produced clock, both movement and case. There is no doubt the movements were mass-produced with interchangeable parts. Nothing I could find explained anything about the cases until I came across mention of the subject in the famous American clockmaker Chauncey Jerome's book, *History of the American Clock Business for the Last Sixty Years*, written in 1860. On page 41 Jerome writes:

"I [Chauncey Jerome] must now go back and give a history of myself; from the winter of 1816, to this time (1825.) As I said before, I went to work for Mr. Terry, making the Patent Shelf Clock in the winter of 1816. Mr. Thomas had been making them for about two years, doing nearly all of the labor on the case by hand. Mr. Terry in the mean time being a great mechanic had many improvements in the way of making the cases. Under his directions I worked a long time at putting up machinery and benches. We had a circular saw, the first one in the town, and which was considered a great curiosity. In the course of the winter he drew another plan of the Pillar Scroll Top Case with great improvements over the one Mr. Thomas was then making. I made the first one of the new style that was ever produced in that factory, which became so celebrated for making the patent case for more than ten years after."*

"When my time was out in the spring, I bought some parts of clocks, mahogany, veneers, etc., and commenced in a small shop, business for myself. I made the case, bought the movements, dials, glass, finishing a few at a time. I found a ready sale for them."

Before going into my interpretation of the above paragraphs, I am struck by the thought of a first-hand account of Seth Thomas and Eli Terry at work with Chauncey Jerome, all giants of American clockmaking.

From the above, clearly Seth Thomas was the first one to make the pillar-&-scroll case. I had assumed it was Terry. Thomas was also making the cases by hand, not mass production. Terry "... had many improvements in the way of making the cases. Under his directions I [Chauncey Jerome] worked a long time at putting up machinery and benches." Knowing Terry and his

**The pillar-&-scroll case was never patented, only the movement.*

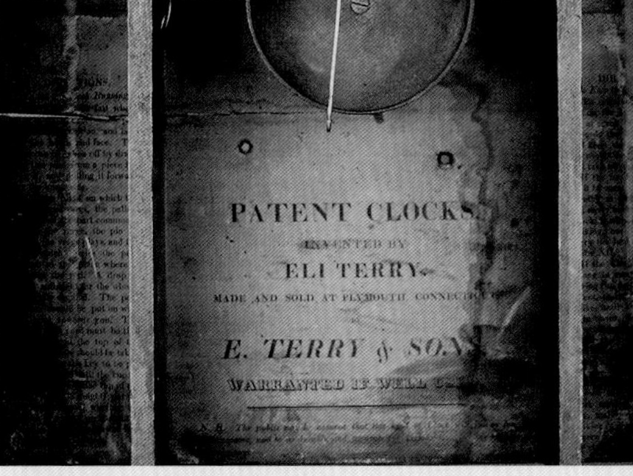

Figure 5.5a (above left). Very rare whale's-tail cased pillar-&-scroll clock made by Eli Terry & Sons, ca. 1829. Eli Terry formed this partnership with his two sons, Eli Junior and Henry in 1823 and it lasted until 1831. Their clocks are sometimes found, as here, with cast pewter hands. Note the beautiful and unusual base. With permission of the clock's owner Christopher R. Brown.

Figure 5.5b (above). Eli Terry & Sons label.

Figure 5.5c (top right). Model 5 movement.

methods for designing machinery to make standardized interchangeable parts for his wooden movements, I feel he was doing the same thing to make the case, as far as possible.

I've included the second Jerome paragraph as an example of what was already happening in the spring of 1817. A single individual with cabinetmaking skills could get up a small operation and make cases by hand, although in this example Jerome had access to case parts from the

Figure 5.6a (above left). The most unusual pillar-&-scroll clock, the "Terry's Deluxe Model," ca. 1818. Note the double C-scroll base. The hands are hand-cut from sheet brass. The six "eyes" in the scroll area are probably ivory. With permission of the clock's owners Cathy and George Goolsby.

Figure 5.6b (above). The Model 3 outside-escapement movement makes this an even rarer find.

Figure 5.6c (left). Eli Terry label.

Terry factory while others would not. They would then purchase the movements from Terry or Thomas and sell the clocks.

I believe the cases made by Eli Terry and Seth Thomas were a standardized product that was mass-produced as much as possible. Terry and Thomas never claimed to have a fully automated process; movements were assembled by hand and it was a time-consuming process. I think the cases were assembled in the factory by workers with some cabinetmaking skills. While Terry's and Thomas's cases were a standardized mass-produced item, many made in small shops that bought their movements probably were not.

Some very rare case variations of the pillar-&-scroll exist and two are pictured, figures 5.5a-5.5c and 5.6a-5.6c.

Terry's and Thomas's production peaked at about 10,000 to 12,000 clocks each in 1825 and the clocks sold for $15. That would be $150,000 to $180,000 each per year in revenue. It would also be about 38 clocks a day leaving each of their factories. In the early years they produced about half that number of clocks per year. Sometimes we have to remember that Seth Thomas and Eli Terry were operating separate factories only a short distance apart and while they worked closely together, they were also competitors with each other.

The pillar-&-scroll was eventually made by many clockmakers. Terry was never able to defend his shelf clock patents, five in all. The clock was a great success through the 1820s. Its arrival signaled the beginning of the end of the wooden-movement tall clock. It was only when more modern and more rugged case styles arrived in the late 1820s that the pillar-&-scroll began to go out of fashion. By 1832 its production was mostly over. Terry's shelf clock movement, made by others and in other case styles, lived on until the mid-1840s.

The significance of the pillar-&-scroll was it was the first mass produced clock—both case and movement—ever produced. This would be the mode of production for all other American clocks to follow.

CHAPTER 6
THE EMPIRE SHELF CLOCK

By 1825 the major clocks one might expect to find in an American home had everything to do with the location of the home and the class of its inhabitants. A fine, established home along the East Coast, such as in Boston, New York, Philadelphia, Baltimore or Charleston, would likely have an eight-day brass-movement tall clock, either American or British, and a French clock or a banjo clock from Boston. No wooden movement clocks here.

Figure 6.1a. Half-column-and-splat wooden-movement shelf clock dating to 1831 to 1832. The splat is a replacement as is the bottom glass which is a mid-nineteenth century print. Courtesy Cowan's Auctions, Inc., Cincinnati, Ohio.

Figure 6.1b. The clock's label shows that the Atkins & Downs' clock was made for George Mitchell. Mitchell, a tavern owner and merchant, was a giant in the Bristol clock industry. He was responsible for bringing many talented clockmakers and casemakers to Bristol. Courtesy Cowan's Auctions, Inc., Cincinnati, Ohio.

Figure 6.2. Bronze looking glass wooden-movement clock. The "bronze" is for the bronze or brass stenciling powder used to decorate the columns and splat. This example has an original reverse-painted glass tablet rather than a looking-glass or mirror in the bottom of the door. Courtesy Cowan's Auctions, Inc., Cincinnati, Ohio.

A better-than-average home in a thriving but mid-sized town anywhere along the settled edge of the frontier such as Natchez, Mississippi; Knoxville, Tennessee; Lexington, Kentucky; or Pittsburgh, Pennsylvania, would likely have an eight-day brass-movement American tall clock and possibly a 30-hour wooden-movement pillar-&-scroll. A log cabin right on the edge of civilization with crops in the ground would make do with a 30-hour wooden-movement tall clock or a pillar-&-scroll. Wage laborers in the East might live with no clock at all.

Clocks were very much a sign of social status. In the whole country people were moving-up. Young adults could rise rapidly either as successful farmers, especially in the South and West, or as merchants and manufacturers all over the country. I have studied the wooden-movement tall-clock manufacturers in the then-thriving state of Ohio where two centers of clock manufacturing were located, and almost all the successful clock manufacturers located there were only in their 20s. Luman Watson of Cincinnati, Ohio, had his own factory when he was just 24, and at that age he had already peddled clocks there for three years and had been in a partnership establishing a factory there for almost another year.

The point to this is that young successful Americans were eager to have the very latest in fashion in their homes. You might live in a log cabin on 300 acres along a fine river—the land equal to any gentleman's estate in Europe—and with the cash from your crops you would love to dress up your meager home with the very latest, most in-fashion clock you could afford. Not the simple wooden-movement 30-hour tall clock you already had, the one with the broken tooth in the movement that no longer ran. You wanted the clock with the greatest flash, the one that the peddler took out of the wagon and proclaimed: "It's all the rage in Boston."

As an aside, I read this above paragraph to a colleague who looked up and said: "Fine story,

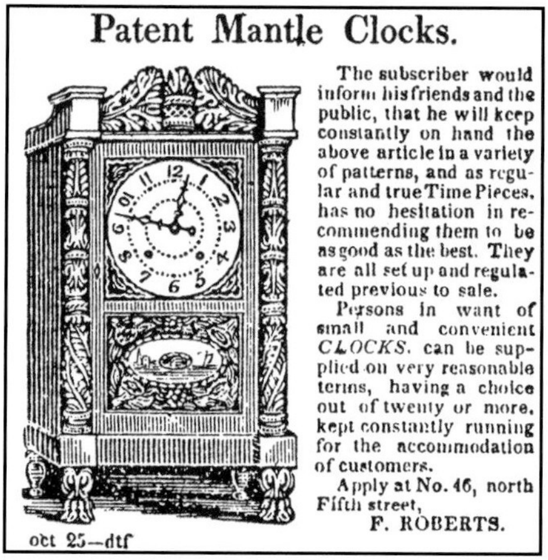

Patent Mantle Clocks.

The subscriber would inform his friends and the public, that he will keep constantly on hand the above article in a variety of patterns, and as regular and true Time Pieces, has no hesitation in recommending them to be as good as the best. They are all set up and regulated previous to sale.

Persons in want of small and convenient *CLOCKS*, can be supplied on very reasonable terms, having a choice out of twenty or more, kept constantly running for the accommodation of customers.

Apply at No. 46, north Fifth street,

F. ROBERTS.

oct 25—dtf

Figure 6.3. Advertisement from Philadelphia, Pennsylvania, newspaper that was first run on October 25, 1827, by clock dealer F. Roberts. Note the similarity with an actual clock of the design by Elias Ingraham for George Mitchell shown in figure 6.4. Reproduced with permission of Dr. Snowden Taylor.

Figure 6.4. Carved column-&-splat wooden-movement clock by Silas Hoadley of Plymouth, Connecticut. This 30-hour shelf clock with the rare upside-down movement was Hoadley's attempt to circumvent the Terry patent. Hoadley simply inverted the patent movement, hanging the pendulum from what would have been the top of the movement. Note: the clock winds from two and ten o'clock above the dial's center-line. Clock from Chris Brown collection. Courtesy Cottone Auctions, Geneseo, New York.

but it was the woman at home to whom the peddler sold the clock while the husband toiled in the field."

The pillar-&-scroll was the first of a rapid series of shelf clocks starting in the 1820s. The 1830s saw a divide between two branches of "in-fashion" clocks. One was a continuation of Terry's or a competitor's 30-hour wooden movement with a dressed-up case. The other was a newer, more expensive shelf clock with an eight-day brass movement. Both new styles were

Figure 6.5. Two eight-day wooden movement clocks. The clock on the right is large, and the one on the left is a giant. While beautiful and collectible clocks, they were expensive and never sold well. Courtesy Cowan's Auctions, Inc., Cincinnati, Ohio.

very popular from 1828 to 1836, with the wooden movement clocks dropping out by the early 1840s and the eight-day brass clocks continuing on with movement modifications.

First we will consider clocks with 30-hour wooden movements and then the eight-day brass clocks.

Shops and factories stayed relatively small throughout the 1830s. A shop, as I understand it, was small and did not have waterpower. There was a major depression in the middle of the decade and that may have slowed the expansion of the factory size for a few years. Canals had arrived and combined with America's great system of rivers, transportation had improved greatly. Manufacturing techniques advanced to where working with brass on a large scale was now possible. The actual technology of making clocks had not advanced considerably past Eli Terry's contributions, but new minor improvements were occurring everywhere. However, Americans still couldn't make springs, at least not coil springs.

The pillar-&-scroll's problem was that it was too successful. It worked well and was reasonably priced. You didn't impress anyone with the purchase of a pillar-&-scroll, especially your wife. What was wanted was an in-fashion clock—but still one that was not too expensive.

Merchants, the interface between the peddlers and the clockmakers, realized that clocks needed to be fashionable before the clockmakers. George Mitchell of Bristol, Connecticut, was such a merchant. Born in 1774, Mitchell—not a clockmaker but a tavern keeper—was probably the man most responsible for making Bristol, Connecticut, the world's leader in clock manufacturing. In 1810 Bristol's population was just 1,428 people.

Mitchell launched a mercantile operation, George Mitchell & Co., just prior to 1809. With the end of the cotton business he brought, among others, cabinetmakers Chauncey Jerome in 1821

Figure 6.6a. Very rare and expensive Joseph Ives "Brooklyn" model shelf clock. Courtesy Skinner, Inc., Boston, Massachusetts.

Figure 6.6b. Wagon-spring powered eight-day movement. This was Ives's first successful attempt at an eight-day rolled-strap-brass movement. He hammered the straight strips of rolled brass into a curve at the top of the movement. The wheels had to be small in diameter and he was forced to use many of them. His later strap-brass movements used thin strips of rolled brass and larger wheels. Courtesy Skinner, Inc., Boston, Massachusetts.

and then Elias Ingraham in 1827 to Bristol. These men made clock cases (not both at the same time) in his old cotton factory. Mitchell purchased movements for these cases from clockmakers Ephraim Downs, Atkins & Downs, Charles Kirk, and others. The clocks he produced had labels such as: "Manufactured by / Atkins & Downs, / for / George Mitchell, / Bristol, Conn." (It is convention to use the "/" character to represent a line break when describing these labels.) The largest type on the above label is "George Mitchell" all in large capitals. Mitchell owned the case factory and was the one that would order the labels and put them in the cases. Mitchell began by making pillar-and-scroll cases but quickly moved to other case styles in keeping with the times. Later he began having others make complete clocks under contract for him with his label, figures 6.1a-6.1b.

Mitchell realized the advantages of moving his clocks and cotton goods to the sea by a canal and was among the organizers of the Farmington Canal Corporation in 1822 and by 1825 work had begun on the canal. His casemaking operation grew and between 1826 and 1828 Mitchell purchased over 3,000 movements from Ephraim Downs. While not the first to make it, Mitchell's first departure from manufacturing the pillar-&-scroll was to make the "bronze looking glass" case. This was a taller, stenciled case (stenciled with bronze and brass powder) and required a 32 rather than the 30-tooth escape wheel on the Terry patent movement. The firm of Jerome, Darrow & Co. first made the bronze looking glass clock back in 1824, but it wasn't popular until Mitchell began mass producing it in about 1828. The Jerome in Jerome, Darrow & Co. was casemaker Chauncey Jerome who had left George Mitchell's employment in 1824. The earlier bronze looking glass clocks had flat columns at the sides, but soon turned columns were used. The turned columns were split in half and each half glued solidly to the case. The

Figure 6.7. Triple-decker shelf clock with eight-day strap brass movement which can just be seen below the 12 on the dial. Courtesy Cowan's Auctions, Inc., Cincinnati, Ohio.

"looking-glass" in the bronze looking glass clock was a mirror. Many of the American clock case styles were taken from earlier mirror designs. The clocks, however, more often had a reverse-painted glass tablet rather than a more expensive mirror, figure 6.2.

At about this same time, late 1827, Mitchell brought cabinetmaker Elias Ingraham to Bristol. Milo L. North, Bristol's foremost historian, wrote in 1872: "Mr. Ingraham was at work at his trade of cabinet making in Hartford [Connecticut], and was recommended by his employer to Mr. Mitchell, who was looking for a man to get up a new style of clock case. This was at the time Chauncey Jerome got up his Bronze pillar case, which was attracting much attention, and Mr. Mitchell's idea was to get up something equal or superior to that. With this object in view, Mr. Ingraham entered into his employ, and soon had accomplished his design, the result being a very handsome case, consisting of carved columns, with lion's paw bases, and fret work heads. Mr. Mitchell purchased his movements from Ephraim Downs [among others]." An advertisement for this new carved clock case appeared as early as October 25, 1827, in Philadelphia, Pennsylvania, figure 6.3. For comparison an actual carved column-and-splat clock is shown in figure 6.4.

This one paragraph quoted above describes the genesis of the two major diversions in style from the pillar-&-scroll. The bronze looking-glass is today known as the half-column-&-splat, or just column-&-splat. The other clock is today's carved-column-&-splat. Largely these two case forms drove the pillar-&-scroll from the market by 1832 and both continued in strong production until the great depression of 1836/1837. The stenciled half-column-&-splat was made by a few manufacturers after the depression but it was largely gone by 1840. Over the years these two basic clocks were produced in a variety of sizes and variations of style in an almost mix-and-match method so that it is possible to find a clock with a carved splat, smooth

Figure 6.8a (opposite). Eight-day brass-movement shelf clock by Birge, Gilbert & Co., ca. 1835. The mirrors are original.

Figure 6.8b. Strap-rolled-brass eight-day roller-pinion movement in Birge, Gilbert & Co. clock shown in figure 6.8a. The movement has pewter winding drums or barrels and even a pewter click.

half columns, and carved feet, or even pillar-&-scroll feet. These mixed cases, while not unknown, are not common.

In 1830 wooden movements from Ephraim Downs cost $3.00 each and cases by Elias Ingraham cost $1.00 to $1.25 each. Henry Terry, Eli's son, used over 4,000 sets of stenciled columns in 1831 alone. Ephraim Downs supplied 7,000 complete clocks (movements and cases) to the trade during the 1830s. He probably acquired the cases by trading for movements. Ephraim Downs was only one of several movement makers in the Bristol area. By the early 1830s the Bristol movement makers began making eight-day wooden-movement shelf clocks. These eight-day clocks required a larger case and sold at a much higher price. They proved difficult to sell and are rare today, figure 6.5. While liked by collectors today, the eight-day clocks were operating at the limit of wooden clock technology. They were not as reliable as the 30-hour models.

For the most part prices for complete wooden-movement clocks held up well during the decade of the 1830s except for the period around the 1836/1837 depression when business failures caused a deep discounting of excess clocks. Wooden-movement clocks always sold for less than their brass cousins, but even at the very end of their run they were not deeply discounted.

In 1843, long after these types of clocks were popular, 30-hour wooden-movement carved half-column-&-splat clocks were still selling for $10 and the stenciled half-column-&-splat was also selling for $10. Eight-day wooden movement clocks sold for $15. Eight-day brass clocks sold for $20. All these prices, with the exception of the depression period, held constant for the decade of the 1830s.

Alongside the 30-hour wooden-movement shelf clocks of the 1830, there was a parallel

Figure 6.9. Salem Bridge eight-day cast-brass movement clock by Spencer, Hotchkiss & Co. of Salem Bridge, Connecticut. The clock has a hand-made cast-brass weight-driven movement. Photograph courtesy Skinner, Inc., Boston, Massachusetts.

development of brass-movement eight-day shelf clocks. The public knew brass clocks were better than wooden clocks. It was just that wooden clocks were easier to manufacture with early nineteenth-century technology than brass clocks. Materials cost less. They were cheap. Gradually, tools increased in quality and brass and iron could be machined successfully.

What Eli Terry was to the development of the wooden-movement clock in Plymouth, Connecticut, Joseph Ives was to the brass-movement shelf clock in Bristol, Connecticut. While Terry was a mechanical genius and a great industrialist, Ives loved to innovate new solutions to overcome problems. The big difference was that Terry was concentrating on producing a good product at a fair, often low price, with an eye towards making a profit. Ives could never focus his efforts to manufacture a product that made him a profit. He could always see shortfalls in the product that needed "improving." He never stayed with anything, producing a handful of this or that and then jumping off into something else. The result was he couldn't make money, was often deep in debt and sometimes in jail for his debts.

Remembering that many people from New England were moving west, the population of Bristol was 1,362 people in 1820, down by 66 people since 1810. It was soon to lose one more as Joseph Ives, deep in debt from unsuccessful clockmaking efforts, departed for Brooklyn, New York. Within a short period of time he declared himself insolvent.

During the period he was in Brooklyn, Ives produced, among other things, a rolled-brass eight-day shelf-clock movement with roller pinions. His first model was powered by wagon springs and it achieved limited production in 1826/1827. The rolled brass was produced in narrow strips. In this first model, the brass for the horseshoe top of the movement was wrought and bent into a curve, figures 6.6a-b. Later movements would use straight sections of rolled brass riveted together. This first movement used relatively wide pieces of rolled brass for the plates and had very small wheels. The later movements used thinner straps and larger wheels. The wheels were also manufactured from rolled brass. This proved to be a successful combination and with further movement modifications, this strap-rolled-brass movement would be the Bristol eight-day brass-movement standard for the decade of the 1830s and beyond. These movements could be powered by wagon springs, remote fusees, or weights.

Ives was in prison in Brooklyn for his debts in 1829. A John Birge of Bristol bailed out Ives and returned him to Bristol. Birge had an eye towards capitalizing on Ives new strap-rolled-brass movement.

The finished Farmington Canal now connected Bristol to New Haven, Connecticut, and thence by river to the sea. This, along with the clock industry, had caused the town to grow. Joseph Ives' uncle, Chauncey, and brother, Lawson C., formed the Bristol firm C. & L. C. Ives in 1829. John Birge, who had bailed out Joseph Ives, was a silent partner in the firm with a 33 percent interest. Joseph Ives worked for the firm, and other Bristol firms, as what today would be a consultant. The strap-rolled-brass eight-day movement soon became the focus of Bristol's production for high-end clocks. Cases for these weight-driven movements were called three-storey and are now known as triple-decker clocks, figure 6.7. They were first made by cabinetmaker Elias Ingraham in 1831 and an entry in his account book for February 1831 reads: "To 104 New Fashioned case at $1.22 - $126.88." Related Bristol firms Birge & Ives, John Birge, and Birge, Case & Co. produced the clocks with C. & L. C. Ives movements. (Note the central figure of John Birge in all four firms. It seems Birge was being paid back for bailing Ives out of jail.)

Later Marsh, Gilbert Co. and George Mitchell also used C. & L. C. Ives movements in their clocks. The movements in these clocks were constantly being modified, which makes the collecting all the more interesting. One major modification was an "A" frame version.

By 1835 there were three Bristol firms producing strap-brass movements. They were C. & L. C. Ives, Birge, Gilbert & Co., and Barnes, Barthelomew & Co., figures 6.8a-6.8b. Barnes, Barthelomew & Co made improvements to the movement that allowed it to run with non-roller pinions. This led the way for the Forrestville Manufacturing Co. and E. C. Brewster—among others—to manufacture movements. Quoting from Ken Roberts's *The Contribution of Joseph Ives to Connecticut Clock Technology, 1860-1862*, page 162: "Spectacular progress in Bristol clockmaking, machinery and techniques literally blossomed during the period 1835/1836. Among new practices were blanking brass movement plates from rolled sheets ... due to the lack of dated documentary evidence as to their origins ... the exact date of these developments ... remains obscure." The Forrestville Manufacturing Co. was the first firm to use blanked plates. By 1836 there were 16 clock factories in Bristol.

There were a few other centers of eight-day brass-movement shelf clock production in America. Several were in upstate New York. One other, and the earliest center, was in Salem Bridge, Connecticut. Heman Clark, the former apprentice of Eli Terry and later Terry's journeyman then partner, was a clockmaker on his own in Plymouth Hollow, Connecticut. He was trained by Terry as both a brass- and wooden-movement maker. Heman Clark left Plymouth Hollow in 1823 and went to work for his brother, Sylvester, at his shop in Salem Bridge, Connecticut. There Heman Clark designed and built an eight-day cast-brass movement for his brother. Two other clockmakers in Salem Bridge, Spencer & Hotchkiss and Spencer, Wooster & Co, also built Clark-style movements. The movements were essentially hand-made and were no competition for the Bristol firms, although Salem Bridge clocks are very collectible, figure 6.9.

As a group the eight-day brass-movement shelf clocks never sold well. They couldn't compete with the cheaper 30-hour wooden-movement shelf clocks which sold for about $10. The brass clocks sold for $15 to over $30.

The depression of 1837, which had started in the West and South, finally made it to New England and ground the clocks of Bristol to a halt. Many firms died. The recovery of Bristol from the "Panic of 1837" was led by the introduction of a new 30-hour rolled-brass movement, an invention of Chauncey Jerome's brother, Noble.

Those interested in buying a 30-hour wooden-movement shelf clock from the 1830s will find they are not too expensive: a couple of hundred dollars. Excellent examples cost much more and are very hard to find. The same is true of eight-day brass clocks, which generally cost about twice as much as the wooden-movement clocks.

CHAPTER 7
THE OG

There were two Jeromes, Chauncey and his brother Noble. Chauncey, the elder, was a cabinetmaker who worked with Eli Terry at Plymouth, Connecticut, in the winter of 1816 to make pillar-&-scroll cases. He moved to Bristol, Connecticut, at the urging of tavern keeper and merchant George Mitchell in 1821 and was on his own making clock cases in Bristol by 1822. At that time Bristol had a population of about 1,400 people and three firms were there making clock movements.

When Chauncey Jerome left Plymouth, he sold his home to Eli Terry for 100 of Terry's shelf-clock movements. Upon arrival in Bristol he purchased 13 acres of land from George Mitchell for the promise of 214 pillar-&-scroll clocks. He finished cases for the 100 movements he had obtained from Terry and got a loan to purchase 114 more movements to allow him to finish the remaining clocks he owed Mitchell.

In Bristol, Chauncey Jerome was reunited with his brother Noble who had been associated with Joseph Ives and had been trained as a brass-movement maker. The brothers were together in the firm Jerome, Darrow & Co. from about 1824 to 1826 and during that time Chauncey Jerome first produced the bronze looking-glass clock using movements by other makers. The bronze looking-glass case—with gradual modifications—became the most popular case design of the 1830s. In late 1826 the firm of Jerome, Thompson & Co. was formed. It lasted less than a year and in late 1827 the firm of Jeromes & Darrow was formed. This was the first time Noble Jerome was elevated to the position of principal owner along with his older brother Chauncey; note the plural Jeromes, for both Chauncey and Noble. This firm was soon to become one of Bristol's major players: in fact, *the* major player. It would be the first time that the Jeromes would produce their own wooden movements, two different models that were both an invention of Noble. The firm of Jeromes & Darrow grew until it was Bristol's largest firm from 1830 to 1833.

In 1830 Bristol had a population of 1,700 people. By 1831 Jeromes & Darrow had added an eight-day wooden movement to their product line. Chauncey and Noble's nephew, Hiram Camp, went to work for the firm. Camp later became superintendent of Chauncey Jerome's movement factory in the early 1840s and then set up a movement factory in New Haven after Chauncey Jerome lost his Bristol factory to fire in 1845. At age 44 Camp became president of the New Haven Clock Co., an industry giant, a position he held until 1893.

One advantage of wooden movements is that they are self lubricating, but in 1833 Jeromes &

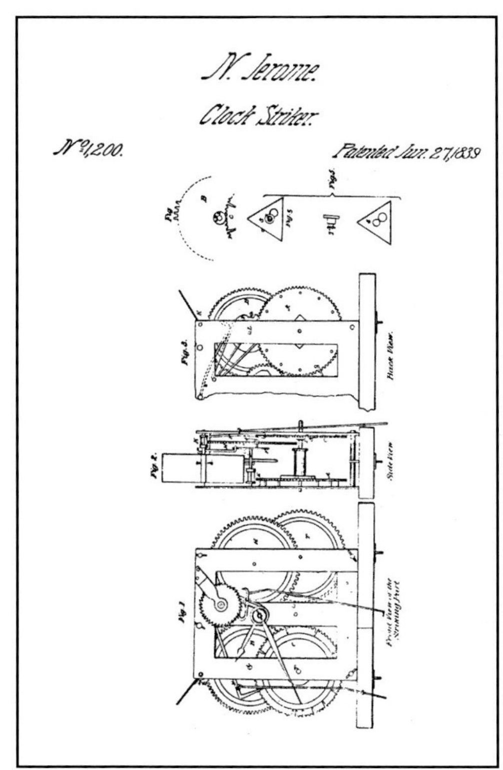

Darrow experimented with brass bushings (bearings) in their wooden shelf clock movements. They looked good, but probably were not an improvement.

The very successful Jeromes formed a new firm, C. & N. Jerome, with a new business plan, in about 1834. Most of the Connecticut shelf clocks were being sold in the South and West and throughout those areas clock peddlers had earned a very poor reputation. States were upset with the peddlers; the fact that they were selling Yankee (northeastern) goods didn't help matters. The southern and some western states required clock peddlers to purchase expensive licenses for each county they sold in and some of these states had 100 counties. This stopped the peddlers: the desired result of the state governments.

An exception was that a license was not required to sell goods made in the state. The Jeromes' plan was to establish an assembly plant and sales office in Richmond, Virginia, the heart of the South, and the clocks they sold would have a label stating they were made in Virginia. By assembling the clocks in Virginia, the Jeromes could attract Virginia merchants and peddlers to buy their clocks because the peddlers would not have to purchase licenses.

It was a success and when Virginia was saturated with Jerome clocks, they repeated the operation in Hamburg, South Carolina, in 1836. Today, clocks with these Southern labels are very collectible and make a premium in the South. They also contain some very unusual brass movements, which we will discuss later.

Back in Connecticut, the Jeromes' Bristol factory, C. & N. Jerome, was producing all the movement and case parts that were being assembled in the South. The labels from Virginia

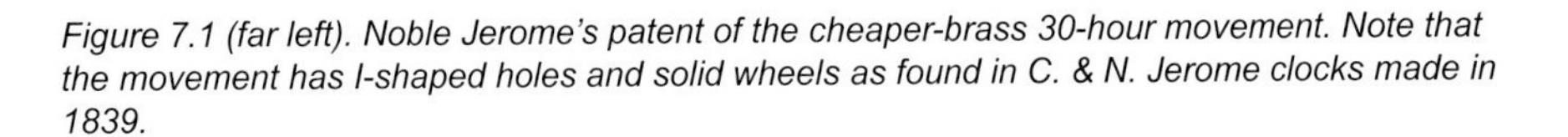

Figure 7.1 (far left). Noble Jerome's patent of the cheaper-brass 30-hour movement. Note that the movement has I-shaped holes and solid wheels as found in C. & N. Jerome clocks made in 1839.

Figure 7.2 (left). OG mirror. Very likely it was this popular mirror that was the idea for the OG clock case.

identified the maker as C. Jerome & Co. and those from South Carolina as L. M. Churchill & Co.

The existence of L. M. Churchill of Hamburg, South Carolina, has only recently come to light. He was a clock merchant born in Oneida County, New York, and married to a lady from Georgia. Again the "& Co." would represent silent partners or partner with some financial stake. Very likely in this instance the "& Co." would have been the Jerome brothers.

The firm of C. & N. Jerome, aware of the success of eight-day brass movements by C. & L. C. Ives, went into the brass-clock-movement business in 1835. Their first effort was an eight-day movement with rack-and-snail striking. It had blanked plates of rolled-brass, non-rolling pinions and was weight-driven. This form of movement was popular in Bristol at the time—and remember, Americans still couldn't make springs. While a good movement, it quickly went out of production at the peak of the depression of 1837. In fact, the whole Jerome empire was on the verge of bankruptcy. In May of 1837 they declared themselves insolvent to the Probate Court and listed as assets 300 finished and 3,000 unfinished brass movements and all the machinery for their production. There were no wooden movements in production so it appears as of this date they were 100 percent invested in brass movements. Somehow, they averted bankruptcy.

The Jeromes knew the marketplace and realised the clock-buying public, just coming out of a depression, would not be anxious to purchase an expensive eight-day brass-movement clock. They might, however, purchase a 30-hour clock for the right price, but the Jeromes had abandoned their capability to produce cheap 30-hour wooden movements. What they needed

Figure 7.3 (above). Typical 30-hour OG clock probably from the 1840s. This style of clock was based on the OG mirror. Photograph courtesy Cowan's Auctions, Inc., Cincinnati, Ohio.

Figure 7.4a (opposite top left). OG clock, the wooden dial signed "C. Jerome" across the top and "Bristol, Conn. / US" at the bottom. The clock's label, however, is "Chauncey Jerome, New Haven, Conn." This clock was likely made in 1845 after Chauncey lost his Bristol factories to fire and was just establishing his New Haven factory. This clock from New Haven was "using up" an existing Bristol dial. The 'U.S.' on the dial usually meant the clock was made for export.

Figure 7.4b (opposite top right). A very similar Chauncey Jerome clock to the one shown in figure 7.4a, this one with a metal dial with the same wording on the dial, but with a Bristol label, meaning this was a slightly earlier clock. Both clocks have the same Fenn stencil on the glass tablet with slightly different color schemes. Mr. Fenn supplied his stenciled glasses to the clock trade starting in 1842. His actual cut stencils still exist at the American Clock & Watch Museum in Bristol, Connecticut. It is very unusual to find two American clocks from the 1840s so much alike. Photograph courtesy Cowan's Auctions, Inc., Cincinnati, Ohio.

Figure 7.4c (opposite bottom left). New Haven label of the clock shown in figure 7.4a.

Figure 7.4d (opposite bottom right). Note that the movement of the clock shown in figure 7.4a has tombstone cut-outs and not the "I" cut-outs of the earlier clocks. It also has spoked wheels. Jerome went to the spoked wheels in about late 1839, and to tombstone cut-outs in about 1843.

was a cheaper brass movement, one they could produce with little capital investment.

The solution was to grind away about the bottom quarter of the blanking dies used to punch out the plates for their eight-day brass movement, thus eliminating the bottom quarter of the movement. The winding drums were moved up to where the second-wheel arbors had been. The train was reduced from four arbors to three arbors and the eight-day movement now ran for 30 hours. The new 30-hour rack-strike movements were assembled in the plant in Virginia and sold there. An alternative view is that they might have blanked out eight-day brass plates and cut off the bottom quarter of the plates, or they might have loaded a short piece of rolled brass onto the long set of dies.

Encouraged by the success of the 30-hour inexpensive brass clock, their next movement advancement was monumental and would affect the American clock market for the next 80 years. The rack-and-snail was replaced with a countwheel strike control and the count plate was regulated by friction drive. This feature was patented by Noble Jerome in 1839 at Hamburg, South Carolina, figure 7.1. Note that this movement has "I" plates, meaning the two holes stamped into the front plate were long, tall rectangles. It also had solid—not spoked—wheels. The first of these movements were made by C. & N. Jerome in 1839 and were often cased in half-round-side cases.

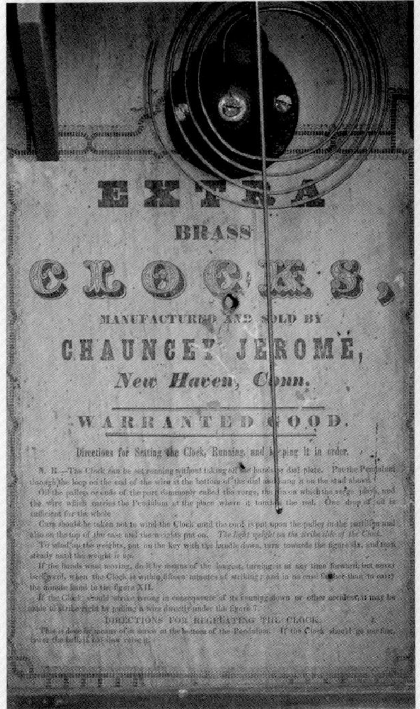
EXTRA
BRASS
CLOCKS,
MANUFACTURED AND SOLD BY
CHAUNCEY JEROME,
New Haven, Conn.
WARRANTED GOOD.
Directions for Setting the Clock, Running, and keeping it in order.
DIRECTIONS FOR REGULATING THE CLOCK.

Figure 7.5a. OG clock made entirely by Chauncey Jerome and shipped to Dayton, Ohio, to be sold there by Buel Pritchard and Joel Holden (working 1841 to 1845) with their label. The printer, Mr. Edwards from New Carlisle, Ohio, spelled their names incorrectly but they used the label anyway.

Figure 7.5b. Label for clock shown in figure 7.5a with names spelled "Prtichard & Holden."

This movement brought the American clock industry out of the depression of 1837 and re-established the Jeromes at the top of the clock industry. Other clockmakers seeing the Jeromes' success were eager to avoid Noble Jerome's patent and produce their own version of the cheaper brass movement. There were many early developmental models of these movements made by many different Bristol clockmakers. Some of their early features are listed here in case a reader just might have one.

- The presence of rack-and-snail striking
- Solid rather than spoked wheels
- Movements with cast-iron back plate
- "Upside down" movement, winding arbors above the center (hand) arbor
- Clocks with a remote spring drive, the springs tucked away in the upper or lower corners of the case
- Clocks with three trains and thus three winding holes

Almost all of the Bristol firms, and Seth Thomas at Plymouth, were soon making the cheaper brass movement. (Note: the term "cheaper" does not mean that the cheaper brass movement cost less than the equivalent wooden movement: the 30-hour wooden-movement clock, complete, was selling at the factory for $4 from 1836 to 1839, while the wholesale price of Jerome's 30-

Figure 7.6a. Another OG clock made by Chauncey Jerome, this one in a beautiful bird's-eye maple case, also labeled by Pritchard & Holden of Dayton.

Figure 7.6b. Mr. Edwards this time spelled their names, again incorrectly, as "Prichard & Holden".

hour cheaper brass movement clocks was $7 initially and $6 later on. It was just that Jerome's 30-hour cheaper brass clock cost less than the eight-day brass clock.)

The early cheaper brass-movement clocks were first found in a case with half-round sides by C. & N. Jerome in 1839, as was mentioned before. This was popular for a while but the mainstay case style for the new 30-hour cheaper brass movement was the OG, sometimes written "ogee." This rectangular-cased clock was first produced by the firm of Jeromes, Gilbert, Grant & Co. starting in 1839 and proved extremely popular. "OG" refers to the "ogive" or S-shaped curve of the wooden molding around the front edge of the case. The Jeromes bought out the other partners in the fall of 1840 and the firm of Jeromes & Co. existed for a couple of months in late 1840. After that, Chauncey Jerome bought out his brother Noble and the firm became just "Chauncey Jerome." As an aside, a cheaper version of the OG clock was made with a wooden movement. Another case style, with a beveled instead of a curved molding, was cheaper still, and it was also usually found with a wooden movement.

The OG case was probably an adaptation of the popular OG mirror, figures 7.2 and 7.3. No attempt was ever made to patent the OG clock case. One great advantage of the brass-movement OG clocks was that they shipped well to overseas markets. Shipments of earlier wooden-movement clocks were made to a few South American countries, but the wooden movements,

Figure 7.7. OG clock with three trains between the plates, the third winding arbor near 6:00 for alarm. The glass tablet is a combination of reverse painting and a mirror, with a beautiful bird at the top of the mirror. Photograph courtesy Cowan's Auctions, Inc., Cincinatti, Ohio.

apparently, didn't stand up well to the saltwater journey. Brass ones did. Soon Americans were shipping the cheaper brass-movement clocks worldwide, figures 7.4a to 7.4d.

In Chauncey Jerome's book, *History of the American Clock Business*, written in 1860, he claims he was the first to introduce the OG clock to England in 1842. It is a great story. However, he might not have been the first. In 1864 J. Leander Bishop wrote *History of American Manufacturers from 1608 to 1860* and on pages 422-423 he claimed that Sperry & Shaw of New York City exported clocks to England in 1841, a year before Jerome.

> *"In 1841, a few clocks were exported there [England] as an experiment. They were seized by the custom house in Liverpool on the ground that they were undervalued. The invoice price was one dollar and 50 cents, and the duties twenty per cent. They, however, were soon released, the owner having accompanied them and satisfied the authorities that they could be made at a profit, even thus low. Mr. Sperry, of the firm Sperry & Shaw, was the gentleman who took out the article. He lost no time, after getting possession of his clocks, in finding an*

Figure 7.8a. OG clock with calendar feature sitting on top of factory-made case. The clock also has an alarm disk under the hands to activate a remote, factory-made alarm. The Connecticut-made clock was modified to add the calendar by, I believe, a clockmaker from near Cleveland, Ohio. In total this clock has time, strike, alarm, and calendar. Photograph courtesy Cowan's Auctions, Inc., Cincinatti, Ohio.

Figure 7.8b. Close-up of calendar feature, a later modification of a Connecticut OG clock. Photograph courtesy Cowan's Auctions, Inc., Cincinatti, Ohio.

auction house. They were made of brass works, cut by machinery out of brass plates, and a neat mahogany case enclosed the time piece... The first invoice sold for four to five pounds sterling, or about twenty dollars each. Since that time every packet [ship] carries out an invoice of the article; and forty-thousand clocks have been sold by this one firm, Sperry & Shaw."

My own research indicates the firm of Sperry & Shaw existed from 1844 to 1851 and that they were merchants not manufacturers in the early years of the firm. They may never have manufactured. One city directory gives the firm's starting date as 1843/44. They advertised in 1846 that they "... made 100,000 clocks a year and had a large export business to England." I have found Sperry & Shaw OG clocks in England. I do not think the firm was in existence in 1841, therefore, Sperry & Shaw could not have made the first shipment in that year. Henry Sperry did go to England, but I do not believe quite as early as 1841. However, I could be wrong.

One last item. In 1841 the OG clock was wholesaling for about $6 and the declared value could not have been as low as $1.50. It is not surprising, if the story is even true, Sperry lied about the value of the clock in the hope they would pass through customs and the 20 per cent import duties would be likewise low.

The population of Bristol, Connecticut, was 2,100 in 1840. The clock factories of Bristol had produced over 1,000,000 clocks in the decade of the 1830s. That would be almost 500 clocks for every man, woman, and child in Bristol.

This is a market that is wide open to today's collectors. Since the 30-hour brass-movement OG was shipped worldwide, it can be found almost everywhere. It is best to collect examples with good glasses, faces, and labels. Restoration of glasses and faces is expensive and does not increase the value in America, but it might be acceptable in other markets. Movement repairs, even if they were not well done originally, do not seem to affect values. Rare early examples are sought-after, but most OGs can be purchased at quite reasonable prices. A good example can be purchased for about $150. Bad clocks cost next to nothing. Examples of the earlier clocks (not OGs) mentioned in this article—especially the clocks from Richmond, Virginia, and even more so clocks from Hamburg, South Carolina—are very difficult to find, as are the clocks with half-round sides. I don't think I have ever seen such a South Carolina clock on the marketplace and I have no idea of the market value. It must be high.

The small place where I live in Ohio—New Carlisle—has a connection with the OG. The firm of Buel Pritchard and Joel Holden, doing business as Pritchard & Holden (1841-1845) in the nearby city of Dayton, Ohio, imported complete clocks from Connecticut and put their label in them. For one such shipment they contracted with the New Carlisle printer C Edwards to print labels for their OG clocks. One batch of labels had one error on a sheet full of printing. It was, however, a serious one, they spelled Pritchard as "Prtichard," figures 7.5a and 7.5b. Pritchard & Holden used the bad labels but had Edwards make a second batch. This time he came closer but spelled Pritchard as Prichard: closer, but not quite right, figures 7.6a and 7.6b. Again Pritchard & Holden used the incorrect labels on their clocks, but they never again seemed to use the services of Mr. Edwards. Edwards' 170-year-old print shop, 20 feet by 90 feet, still stands and is full of trays of type and old printing machines, but is no longer in use.

In summary, the OG clock is important but it is also one of the most overlooked by collectors. It shouldn't be.

Wooden nutmegs and cucumber seeds

Chauncey Jerome, late in his life and looking back at his accomplishments with both the wooden and brass movement clocks, commented, "Wood clocks were good for a time, but it was a slow job to properly make them, and difficult to procure wood just right for the wheels and plates, and it took a whole year to season it ... they were always classed with wooden nutmegs and wooden cucumber seeds." What were wooden nutmegs and cucumber seeds?

Lads in the clock shops, hired to sweep up after a day of turning and sawing, would take scraps of wood and turn out wooden nutmegs and carve wooden pumpkin seeds. The lads would sell these to unscrupulous peddlers who along with the clocks, would sell them to the farmers. I have examined lawsuits about clock peddlers selling wooden nutmegs and cucumber seeds to farmers. It was considered a serious crime and paraphrasing a country judge: "It is a triply serious crime to sell bad seed as it takes money for worthless goods and it takes the time of the farmer's wife to plant them and possibly the lives of the whole family if they starve for the lack of food!"

CHAPTER 8
THE SPRING CLOCK

Until now—the 1830s—weights had been the normal way of powering American clocks. As we have seen, Joseph Ives had experimented with leaf springs, or wagon springs, to power his clocks, and while they worked, they were not economically successful. This is the first time we will discuss American clocks with coil springs.

Why couldn't Americans make springs? The answer is that they couldn't make high carbon steel. The Englishman Benjamin Huntsman began making crucible steel in quantity near Sheffield in 1740. This became known as English cast steel and was used for files, edge tools, clock pinions, and clock springs. America was 50 years or more behind Britain in the production of cast crucible steel. The Americans couldn't make the high carbon steel and they couldn't afford to import the expensive clock springs from Britain or France. If Americans could make clocks out of wood, possibly they could make springs out of something other than steel. They did.

Joseph Shaylor Ives, the nephew of Joseph Ives, first appeared in Bristol, Connecticut in 1832. He worked for Elisha C. Brewster as a clock mechanic and in May of 1836, Joseph Shaylor Ives obtained a patent for brass mainsprings for clocks. Brewster, eager to produce spring clocks, purchased the rights to Joseph Shaylor Ives' patent in exchange for a house. This was the beginning of mass-produced spring clocks in America.

Elisha C. Brewster was trained as a tailor and came to Bristol in 1819 at the age of 28 to work as a clock peddler in the South. Ten years later he was operating a small dial shop in Bristol, but it seems as if he never worked at any mechanical aspect of clock manufacturing. Brewster was associated with Charles Kirk who, like Joseph Ives, was a mechanical genius who was always in financial difficulties. Among others, Charles Kirk was in debt to E. C. Brewster.

Brewster cleared Kirk's debts by purchasing his clock shop in 1833 and he kept Kirk on as an employee to manage his new business of manufacturing brass-movement shelf clocks. By the mid-1830s Brewster was producing an eight-day weight-driven rack-and-snail rolled-brass movement with fixed lantern pinions as other Bristol firms were manufacturing, especially C. & N. Jerome. By 1836 Brewster, now employing Joseph Shaylor Ives as well as Charles Kirk, had converted this movement to spring drive by adding remote fusees and installed it in a round Gothic case, figures 8.1a, 8.1b and 8.1c. Today this clock style is called a beehive clock, figure 8.2.

The American fusee system is different from the British. The British placed the spring in a cylindrical spring barrel, around the smooth outside of which they wrapped the gut or chain. The fusee was on a separate winding arbor. The Americans, possibly because the first application was by Charles Kirk to an already-existing eight-day movement, placed the spring inside the fusee or attached to a fusee. They then wrapped cord around the fusee which in turn drove the existing smooth barrel on the winding arbor.

Early applications had the fusee remote, not actually attached to the movement, again because

Figure 8.1a. Shelf clock from about 1839 attributed to E. C. Brewster of Bristol, Connecticut. Photograph courtesy of Skinner, Inc., Boston, Massachusetts.

Figure 8.1b. Eight-day remote-fusee rack-striking movement by E. C. Brewster in clock shown in figure 8.1a The remote fusee uses the American system where the fusee is connected to the spring while the winding arbor has the smooth barrel which the cord wraps around. Photograph courtesy of Skinner, Inc., Boston, Massachusetts.

existing mass-produced weight-driven movements were being used. Later on the fusee was an integral part of the movement and still later the fusee was abandoned and movements were designed that had direct-drive internal springs. At first springs were brass, and later steel. All this development happened in just over ten years.

Remembering the American financial situation, 1836 could not have been a worse time for E. C. Brewster to attempt to place his fusee beehive clock on the market. Within a year the country was deep into the depression of 1837. Sales very likely lagged and Brewster took Joseph Shaylor Ives and Charles Kirk into his firm as partners and the firm's name changed to E. C. Brewster & Co. in 1839. The firm's next advancement, no doubt an improvement of Charles Kirk's, was to produce a rack-strike, eight-day movement with a cast-iron back plate which included iron spring containers and eliminated the fusees. The cast-iron cup that contained the spring was probably thought necessary to contain the destructive power of a breaking spring. It didn't prove necessary. This movement could have been in use as early as 1838. Again, it was mostly offered in a beehive case. Charles Kirk also designed a 30-hour version of this cast-iron

Figure 8.1c. Closeup of the rare rack in this Brewster movement. Photograph courtesy of Skinner, Inc., Boston, Massachusetts.

Figure 8.2. Gothic or beehive clock by Brewster & Ingrahams, a case style that most of the early Brewster movements are associated with. This case style was an innovation of Elias Ingraham. Photograph courtesy of Skinner, Inc., Boston, Massachusetts.

back plate movement with count-wheel control. These clocks are very collectible and expensive and often found in England and Wales.

E. C. Brewster & Co. often used a small card label to identify their clocks, many of which clocks have lost their little labels which were about the size of today's business card, but the clocks usually had colored paper lining the backboard and often the tacks that held the card or the holes from the tacks are still visible in the paper. Most of these movements are stamped with the firm's name (as were all of Brewster's early movements), and some of the card labels were printed as "E. C. Brewster & Co. USA," the "USA" indicating export, most likely to Britain.

In 1843 Brewster merged his firm with the two Ingraham brothers, Elias and Andrew. Elias had been largely unemployed and Andrew had been in the casemaking firm of Ray & Ingraham. The new firm was Brewster & Ingrahams; the "s" on Ingraham was for the two Ingrahams, Elias and Andrew. Brewster & Ingrahams were major exporters to Britain. In 1845 they changed the back plate from cast iron to brass but initially retained the cast-iron spring retainers because they were still thought necessary.

Figure 8.3. Sharp Gothic case, another case first produced by Elias Ingraham. This clock carries the label of Chauncey Jerome working in Austin, Illinois, just before his death in 1868. Photograph courtesy of Skinner, Inc., Boston, Massachusetts.

Figure 8.4. Four-column steeple or sometimes twin-spire steeple by Brewster & Ingrahams also designed by Elias Ingraham. Photograph courtesy of Skinner, Inc., Boston, Massachusetts.

Clocks that have any portion of the movement cast iron are very collectible and therefore valuable.

Elias C. Ingraham, the cabinetmaker and partner in the firm of Brewster & Ingrahams, was responsible for many of the spring-clock case designs. He conceived the beehive (round Gothic) case, figure 8.2, as well as the steeple (sharp Gothic), figure 8.3. He later modified the sharp Gothic to the four-column steeple and likewise designed the four-column OG beehive, figures 8.4 and 8.5. All these clocks proved very popular in Britain and Brewster & Ingrahams had both a London office, 13 Walbrook, and a Liverpool office. The firm of Brewster & Ingrahams dissolved in 1852 when E. C. Brewster left the company. The firms E. & A. Ingraham & Co. and the Brewster Mfg. Co. were formed in 1852. Both offered a new case design of a small scroll-front shelf clock often with a "ripple" door, figure 8.6. The Brewster Mfg. Co. lasted until 1854 and E. & A. Ingraham & Co. dissolved as a result of another American depression in 1855 and a fire in 1856.

Back in 1839, Chauncey and Noble Jerome were ready to go into large-scale production of their cheaper brass movement in an OG case, but they needed additional capital and took two major partners, William L. Gilbert and Zelotis Grant, into their partnership, dissolving C. & N. Jerome and forming Jeromes, Gilbert, Grant & Co. It was this firm that first mass produced the weight-driven 30-hour cheaper brass movement in the OG case. This firm lasted from 1839 until 1840 when Chauncey and Noble Jerome bought out the others. Shortly after that Chauncey brought out his brother Noble and began trading in his own name, Chauncey Jerome, as a wholly owned firm at the start of 1841.

Flushed with the success of the OG clock, Chauncey Jerome established Jerome's Clock

Figure 8.5. Four-column OG Gothic or OG beehive, yet another case style by Elias Ingraham. This clock is by the Brewster Manufacturing Co. Photograph courtesy of Skinner, Inc., Boston, Massachusetts.

Figure 8.6. Brewster Mfg. Co. small eight-day stenciled shelf clock, ca. 1853, with ripple molding on door. E. & A. Ingrahams made the same ripple door clock. Ripple molding clocks are very collectible.

Store in New York City to sell his clocks. He successfully penetrated the British market in 1842 and in 1843 he formed and was president of the Bristol Clock Co. to sell clocks through agents in China. The clocks were rolling out, money rolling in.

Between 1843 and 1845 Chauncey Jerome was the largest manufacturer of clocks in Bristol, in Connecticut, in the United States, and in the entire world. But in 1845 the whole business literally went up in smoke. Jerome had two Bristol factories and a New Haven, Connecticut, case factory. The two Bristol plants were destroyed by fire. In severe financial trouble, Chauncey Jerome gave up his firm and went into partnership with his nephew, Hiram Camp and formed the Jerome Mfg. Co. in New Haven. While the firm was legally the Jerome Mfg. Co. in New Haven, the clocks they produced there were mostly labeled "Chauncey Jerome" to capitalize on his good name. Immediately after Jerome's fire, Brewster & Ingrahams was the largest manufacturer of clocks in the world for a little while until the Jerome Mfg Co got on its feet.

Joseph Ives continued to refine his wagon spring until 1845 when he patented a constant force example in both 30-hour and eight-day versions, figure 8.7. This wagon spring coupled with its intermediate drums provided consistent force to the movement. It was a success and the firm of Birge & Fuller had sole use of the Joseph Ives patent and is noted for his wagon springs in their clocks, figures 8.8, 8.9 and 8.10.

In 1846 two firms controlled the American spring-clock market. Birge & Fuller used Joseph Ives' wagon springs and Brewster & Ingrahams used Joseph Shaylor Ives' brass coil springs. The competitors were shut out—until Edward L. Dunbar was successful in producing steel coil springs in 1847. Silas B. Terry, the youngest son of the famous clockmaker Eli Terry, is credited

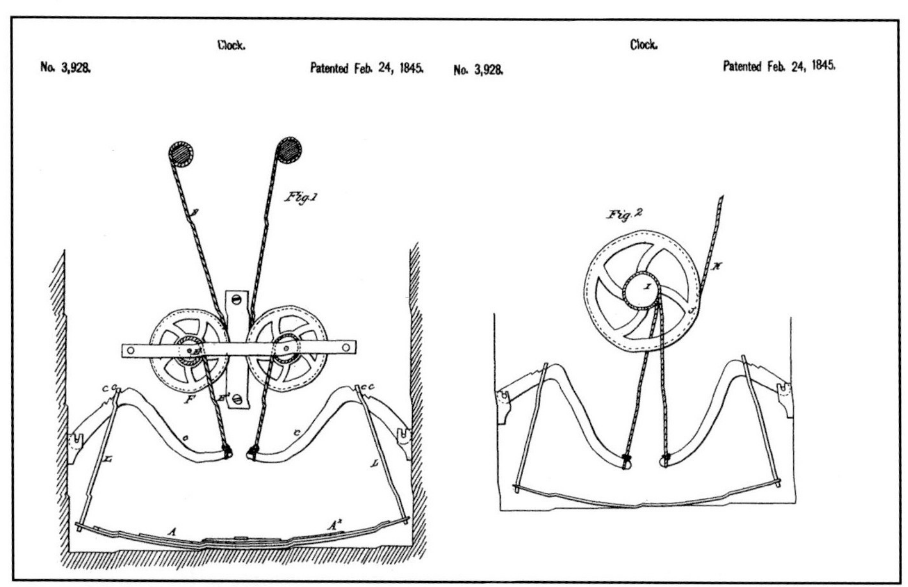
Clock.
No. 3,928.
Patented Feb. 24, 1845.
Fig.1
Clock.
No. 3,928.
Patented Feb. 24, 1845.
Fig. 2

Figure 8.7 (opposite). Joseph Ives' patent drawing for constant-force wagon spring of 1845. These wagon springs are often found in Birge & Fuller clocks with Birge & Fuller-made movements. The drawing on the left depicts an eight-day wagon spring clock and the one on the right a 30-hour.

Figure 8.8 (opposite bottom left). Double-candlestick clock with Birge & Fuller movement with remote wooden fusees. These cases often have Ives wagon springs, but not this one. Photograph courtesy of Skinner, Inc., Boston, Massachusetts.

Figure 8.9 (opposite bottom right). Double-candlestick clock by Birge & Fuller, ca. 1845, with Ives eight-day patent wagon springs. Photograph courtesy of Skinner, Inc., Boston, Massachusetts.

Figure 8.10. Single candlestick beehive 30-hour wagon spring clock by Birge & Fuller, ca. 1850. Photograph courtesy of Skinner, Inc., Boston, Massachusetts.

with the development of an American process for producing steel coiled springs. From an 1872 article by M. L. Norton looking back at the process:

> *"Mr. [Silas B.] Terry claimed to be the inventor of the process which was by means of heated tallow and was designed to cheapen the manufacture of springs; as they had formerly been purchased of French makers at one or more dollars apiece. Mr. John Pomeroy experimented with tempering springs and probably succeeded by another process at the same time with Mr. Terry. His process consisted, in part, of the use of melted lead and immersing in oil, being first heated in an oven over a wood blaze. Terry's process heated the springs directly over a coal fire, and by dipping them in tallow, as before mentioned."*

Silas B. Terry sold his secret of making steel springs in 1847: "In consideration of five hundred dollars to me paid, I hereby promise to impart to Edward L. Dunbar, Lorenzo D. Jacobs, Winthrop Warner all the information I possess by telling and showing them how to harden and temper clock springs to sell to all persons who wish to purchase the same, *except Chauncey Jerome of New Haven*" [my italics]. Once more Chauncey Jerome, the giant of the industry, was shut out of the spring clock business.

Thomas Fuller—the Fuller in Birge & Fuller—died in 1849, and Birge continued on as John Birge & Co. This company had an export business to Britain, and many of its very interesting clocks have been found there.

By 1848 Americans were making steel spring clocks, and by 1850 brass springs were a thing of the past, except for Seth Thomas who continued to use them into the 1850s. Even Jerome had found a way to get steel springs for his clocks. Jerome was very successful with his business

Figure 8.11a (above). Drop octagon eight-day fusee dial clock by Chauncey Jerome, ca. 1850. Other Bristol clockmakers tried to keep Jerome out of the spring clock business but he had spring clocks by 1850. Photograph courtesy of Skinner, Inc., Boston, Massachusetts.

Figure 8.11b (above right). The ca. 1850 eight-day movement is stamped "Chauncey Jerome, New Haven, Conn. USA" indicating it may have been destined for the export market. The detached or remote fusees are of the American type with the spring attached to the fusee. Photograph courtesy of Skinner, Inc., Boston, Massachusetts.

Figure 8.12a (opposite top left). Medium-size cottage clock by Chauncey Jerome of New Haven, Connecticut, from the early 1850s This 30-hour example is in a bird's-eye maple case. Photograph courtesy of Skinner, Inc., Boston, Massachusetts.

Figure 8.12b (opposite top right). A 30-hour remote fusee movement with brass fusee bracket. Photograph courtesy of Skinner, Inc., Boston, Massachusetts.

Figure 8.13a (opposite bottom left). Small eight-day Chauncey Jerome, New Haven, Connecticut, cottage clock from the early 1850s. Courtesy of Skinner, Inc., Boston, Massachusetts.

Figure 8.13b (opposite bottom right). Eight-day remote fusee movement with cast-iron frame to hold fusees. Photograph courtesy of Skinner, Inc., Boston, Massachusetts.

GOTHIC
WITHOUT ALARMS,
SPRING
Sharp & Round Top,

Figure 8.14a. Chauncey Jerome, New Haven, Connecticut, giant steeple. It is about half again the size of a standard steeple. This example is veneered in kingwood.

Figure 8.14b. Eight-day fusee movement of Jerome giant steeple. While it looks like a fusee clock, it has no fusees and is simply a remote-spring clock. Photographs courtesy of Doug Cowan.

in Britain, called Jerome & Co. He quickly found out that the British clockmakers found fault with his steel spring clocks, because they did not have a fusee and he hastily incorporated fusees into his clocks, figures 8.11 to 8.13. One of his fusee spring clocks was a giant steeple. In fact, it was not a fusee clock but rather a remote spring clock. It looked like a fusee clock, but it had no fusees, figure 8.14.

Seth Thomas of Plymouth, Connecticut, had a habit of not experimenting the way the Bristol makers did. He waited until a movement model or a case style proved itself with the public and then made the same thing, but even better than the originator, and stayed with it for a long time. Therefore it not unusual to find Seth Thomas clocks that are made later than you might expect them to be. He was always slow to the marketplace.

This pretty much concludes the story of the introduction of springs into American clocks, but the Americans never abandoned weight clocks. The 30-hour weight-driven OG was still being offered during World War I!

In 1850 the population of Bristol, Connecticut, was 2,880 people and amazingly for the first time it was just slightly larger than it had been in 1800. The great exodus from Connecticut to the West started in 1800 and within 10 years about half the people had left the state; many went to New York State and Ohio.

Collectors of American clocks will find that it is not difficult to find clocks with brass springs. A lot of owners don't know they are in their clocks, but repairmen often find them. As I mentioned in an earlier chapter, if a broken brass spring needs to be replaced, keep the broken spring with the clock. Also, brass springs are not as powerful as steel, so do not replace a brass spring with the same size steel spring; it will be way too strong. Finding clocks with cast-iron movement parts, fusees, or wagon springs will take a bit of luck, although they are not overly expensive.

CHAPTER 9
THE MAGNIFICENT SEVEN

The first chapter of this book was an overview of American clocks and the next seven chapters chronologically presented American clocks from their colonial beginnings to the period just before the advent of the major American clock manufacturers. These large firms would dominate American clockmaking—and thus the clocks themselves—through the later half of the nineteenth century and into the twentieth.

By the end of the Civil War, 1865—and just after—just as the larger clock factories from Connecticut began experiencing explosive growth, many factors came together, making the continued rapid growth of the clock industry possible. It was a sort of perfect storm that would shower America with clocks. The railroads crossed the nation all the way from the East Coast to the West. Along with the advanced railways came the telegraph and government postal systems that took advantage of the railways, the telegraph lines running along the railroad right-of-way, and the freight and letters traveling on the trains.

The clock companies published their own catalogs of their goods and often general goods catalogs had a section on clocks that contained models from each of the major clock companies. This meant that even the smallest general store in the most remote corner of the new West not only had a clock or two on their counter, but access to page after page of clocks in their store catalogs. If a clock could be imagined, it could be had, anywhere in America.

No longer would the peddler show up on the farm once a year with three or four clocks, like they did a generation or two before; now every week or so the tug of a new clock pulled at the family as they bought their beans, flour, and lard. It is not surprising that the clock companies responded with all types of clocks at all sorts of prices: the combination of those two factors would eventually come together to provide everyone with just the right clock at just the right price.

There were seven major producers of American clocks all located in the tiny state of Connecticut. In this last chapter I present each of those seven firms one at a time from its beginning to about 1900. Along the way I discuss some of the clocks made by the company as well as the company itself and with each I select what I think is a key clock for more in-depth examination.

The seven firms are Seth Thomas, New Haven, Ingraham, Ansonia, Waterbury, Gilbert and Welch/Sessions. Admittedly, this approach excludes hundreds of smaller firms and a few large ones. For example, the alarm clock manufacturers of the Midwest made vast numbers of clocks, most of them in the twenthieth century, and are outside the scope of this book. Calendar clocks were made by many firms in New York State but not in large numbers, compared to the seven giants from Connecticut. Likewise, the higher-quality firms of the Boston, Massachusetts, area made fantastic clocks but not many of them, especially not in the nineteenth century.

Seth Thomas

If you look back to the names of the seven companies, you will realize that three are named for a location and the rest for a person. The Seth Thomas Clock Company cannot be discussed without mentioning Mr. Thomas himself.

In 1853 Seth Thomas, at the age of 67, a giant of the clock business, and "… feeling the infirmities of the years coming upon him …" decided to form the Seth Thomas Clock Company with a capital value of $75,000. Most of the other key individuals in the clock business were associated with Bristol, Connecticut, but Seth Thomas was a Plymouth, Connecticut, man, so much so that shortly after his death in 1859 people began calling the place Thomaston in his honor. By 1865 it was Thomaston on the clock labels, and by 1875, Thomaston by law. One date you might want to remember is that Seth Thomas clock labels identified the place as Plymouth or Plymouth Hollow before 1865 and from 1865 on as Thomaston. To put that date into context, the Civil War ended in 1865.

Seth Thomas was conservative. Seth Thomas the company—long after the man's death—was still conservative, not an innovator, not willing to push rapid changes to the marketplace, not willing to take chances, always taking the time to design and build a better clock than the others, and slow to let go of a product. I find this last trait interesting and have never heard it mentioned in any of my business training. An old-fashioned clock, an old standard, would appeal to a segment of the buyers and when you are the only firm producing that clock, it must have resulted in

Figure 9.1a (opposite page). My Seth Thomas Regulator Number 2 is of mahogany or cherry. It is 3ft tall with a 12 inch dial. Everything about the clock is completely original including the deep red-brown finish. The dials of many Seth Thomas clocks are prone to flaking. It is always a question as to when such a dial should be repainted, but this clock's dial is untouched. This clock has a coded date stamp on its back and it dates to about 1890.

Figure 9.1b (left). The clock shown with the top and bottom doors open.

Figure 9.1c (bottom). One of the really great things about this clock is that it still retains its setup instructions. These are almost always missing, serving no purpose after the clock is up and running.

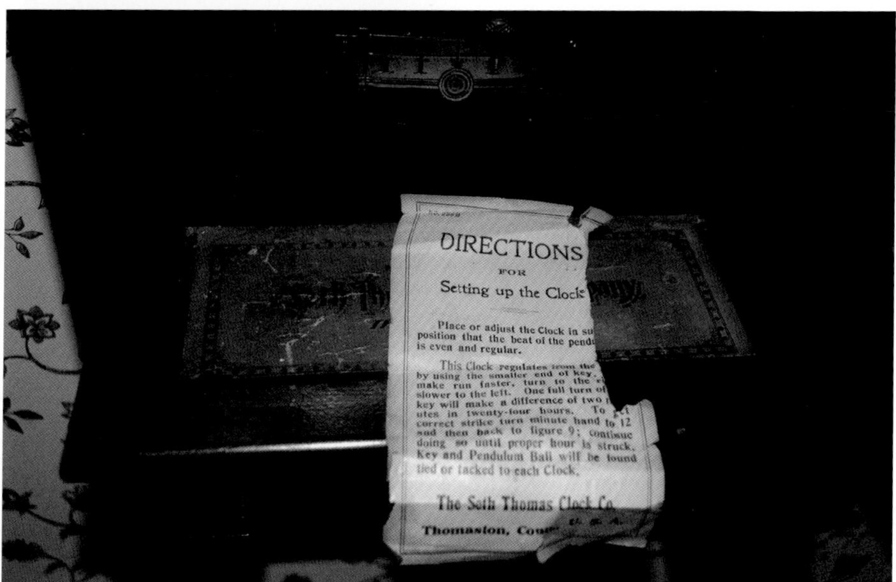

DIRECTIONS

FOR

Setting up the Clock

Place or adjust the Clock in su
position that the beat of the pendu
is even and regular.

This Clock regulates from the
by using the smaller end of key.
make run faster, turn to the
slower to the left. One full turn of
key will make a difference of two
utes in twenty-four hours. To get
correct strike turn minute hand to 12
and then back to figure 9; continue
doing so until proper hour is struck.
Key and Pendulum Ball will be found
tied or tacked to each Clock.

The Seth Thomas Clock Co.

Thomaston, Conn U. S. A.

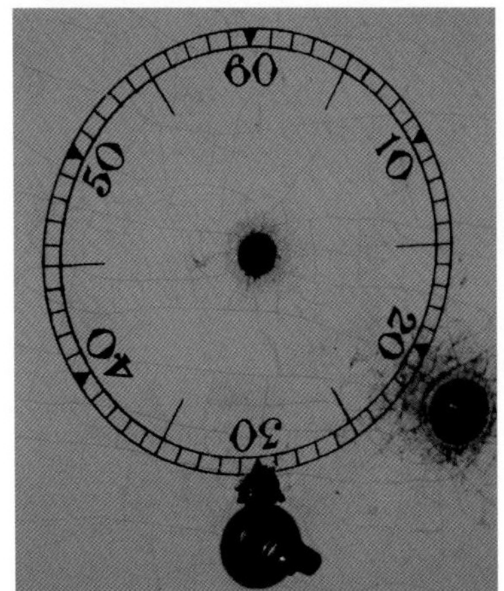

Figure 9.1d (top). The clock's label is located beneath the pendulum. The meager crank is original. The clock has a beat regulator.

Figure 9.1e (above left). When removing the dial, there is an additional almost hidden screw under the seconds bit. Many of these clocks have damage to their zinc dial when repairmen or collectors remove all the screws around the dial and attempt to remove the dial by pulling it off when this small but powerful screw is still holding fast.

Figure 9.1f (above right). The stick across the front of the movement is only there to hold that small screw shown in figure 9.1e. Many clocks have had this stick removed to eliminate the trouble of having to remove it and reinstall it to service the movement. It is nice to have this stick still in place.

adequate sales to keep it in production.

The ultimate example is the steeple clock, a design of Elias Ingraham in the early 1840s. It was all the rage with the Bristol clockmakers in the late 1840s and 1850s, slowing down in the 1860s. The Seth Thomas Clock Co. introduced their steeple clock in 1879, almost 40 years after Ingraham designed it, and produced it until 1917. That's conservative!

Seth Thomas's youngest son Aaron became president of the Seth Thomas Clock Co. nine days before Seth Thomas died. During the 1860s the firm prospered. In 1860 it started

Figure 9.1g (above left). The regulator movement has maintaining power on the great wheel.

Figure 9.1h (above). The left-side view of the movement shows it has open lantern pinions.

Figure 9.1i (left). Just to the right of the hand arbor is the Seth Thomas maker's stamp and "Made in US America." The Seth Thomas No. 2 movements have been reproduced in America and India, and possibly also China. None of the reproduction movements that I am aware of have maker's stamps. A highly original clock such as this would cost a great deal.

making regulators and by 1863 had three of them in production. The No. 2 regulator was very successful, and it was produced for over 90 years when it was dropped from production in the 1950s, only to be revived by the company in 1976 as a "limited" production of 4,000 clocks.

An offshoot brand of the firm was Seth Thomas' Sons & Co. which was formed in 1865. This company made clocks in imitation of French clocks, but not quite of as high quality. It flourished for a while but by 1879 was taken over by the parent company, Seth Thomas.

The Seth Thomas Clock Co. had sales offices in New York, Chicago, and San Francisco by 1868. In 1874 they opened a London office and in 1890 they added a St. Louis, Missouri, office. In 1884 Seth Thomas began manufacturing watches, but by 1915 the production of watches had ceased.

Seth Thomas produced a line of clocks much like the other Connecticut firms. One of the interesting lines of Seth Thomas clocks for collectors is the city series. I don't believe they conceived the clocks as a series; they just produced high-grade walnut parlor clocks with names of US cities. The city series clocks often had a unique pendulum for each model, or pendulums that were only used for a very small number of the city series models. The clocks were produced for over a decade with new city series models constantly coming on the market. Many collectors find these clocks make an enjoyable collection. They are not overly expensive and, like all Seth Thomas clocks, are of very good quality.

One other aspect of Seth Thomas clocks is that they often have a date code, usually stamped in black ink on the back of the case in about 5/8 inch high numbers preceded with a letter, such as G3981. The first letter is the month of manufacture with January being A, February B, etc. The four numbers are the date in reverse order. In our example G, the seventh letter of the alphabet, stands for the seventh month, July, and 3981 reversed is 1893. So our example would have been made in July of 1893.

One of the few inventions of the firm was the process of producing a celluloid plastic, "Adamantine," which was patented in 1885 and proved very popular until 1917 when it went out of production. They used Adamantine in many colors to cover many of their flat-top mantel clocks.

In 2001 the firm filed for bankruptcy and the Colibri Group of Providence, Rhode Island, purchased the rights to the Seth Thomas name; and Seth Thomas clocks, probably produced in Asia, are being sold today, 2011.

My favorite Seth Thomas clock is the Regulator No. 2 and my own example is shown in figures 9.1a-9.1i.

New Haven

New Haven is a port city located at the mouth of the Connecticut River about 60 miles south of the clock-producing town of Bristol. By the late 1830s the Farmington Canal connected the two places and in 1844 Chauncey Jerome built a casemaking factory in New Haven and shipped movements from his two movement making factories in Bristol to be cased in New Haven. The finished clocks were shipped from New Haven to the rest of North America and around the world. Between 1843 and 1845 the firm of Chauncey Jerome was the largest manufacturer of clocks in the world.

Figure 9.2a. This 30-hour OG clock dates to about the 1870s. The clock was made by the New Haven Clock Co. but contains the Jerome & Co. label, indicating it was made for sale in Britain. The bottom glass, or tablet, was of a style common to many New Haven clocks of this period. Thanks to Earl Harlamert of Dayton, Ohio, for allowing his clock to be photographed and thanks also to Tom & Fran Davidson of New Carlisle, Ohio, for allowing their new home to be used as a set for the photographs.

Figure 9.2b. The interior of the clock shows the "Jerome & Co. New Haven, Conn" label.

Disaster struck in 1845 when fire destroyed Jerome's movement making factories in Bristol. Chauncey Jerome's nephew, Hiram Camp, had been the superintendent of the Bristol movement factories and Jerome had sent him to New Haven in 1844 to construct the casemaking factory there. After the fire Camp supervised the addition of a movement factory to the case factory in New Haven, and four months after the fire, Chauncey Jerome was back in business this time wholly in New Haven. Clocks made in New Haven went out with Chauncey Jerome, New Haven, labels

In 1847 Camp was independently operating his own clock factory on a small scale delivering complete clocks, boxed for shipment by Jerome. In 1850 Chauncey Jerome and several members of the board of directors of the brass manufacturers Benedict & Burnham formed a joint stock company with half the shares owned by Chauncey Jerome. The firm was the Jerome Manufacturing Co. By 1851 Camp had built a wooden factory near the Jerome Manufacturing

Figure 9.2c. The interior with the dial removed and pendulum installed. The two iron weights, 2½ pounds each, run down the two channels to the left and right of the movement. The movement mounts on a wooden seat board, which slides into slots in the vertical wooden stiffeners. There are two pulleys in the top of the case for the weight cords to run over.

Figure 9.2d (opposite). A closer view of the movement installation. The gong mounts under the movement. At the top of the interior one can see a hole in the back of the case. This "shelf" clock was converted by an early owner to be a wall clock, something that happened often in Britain where this clock probably lived for its first 100 years. I have never seen an OG sold new in America so modified to be a wall clock. Personally, I think they do well as wall clocks because they are somewhat unstable with the weights fully wound to the top of the case when sitting on a shelf. If hung on the wall, on the other hand, when the weight cord breaks the weight will take out the bottom of the case.

Co. in New Haven to build movements for Jerome. Step by step Hiram Camp was becoming independent of Jerome.

In 1853 Camp's wooden movement factory was incorporated as the New Haven Clock Co. with Hiram Camp as president to "manufacture, sell and deal in clocks and timekeepers of every description." Camp was a 25 percent owner. The fledging company was a major supplier, mainly of movements, to the Jerome Manufacturing Co.

Financial disaster struck Chauncey Jerome in 1855 and by the following year the Jerome Manufacturing Co. was bankrupt. Chauncey Jerome was destroyed, but the New Haven Clock Co. continued and bought the assets of the failed Jerome Manufacturing Co. Camp continued on as president of the New Haven Clock Co., a position he held until January 1, 1892. With Chauncey Jerome out of the picture, the New Haven Clock Co.'s largest customer became Jerome & Co. of Liverpool, England, and they made all the clocks for that company until the twentieth century.

The New Haven Clock Co. was successful during the 1860s and 1870s selling its line of

Figure 9.2e. A close-up view of the New Haven 30-hour brass movement. The president of the New Haven Clock Co. was Hiram Camp in the 1870s when this clock was made, and it was Hiram Camp who earlier ran the Jerome movement factories for Chauncey Jerome in Bristol prior to 1845 and in New Haven after that date. Camp formed the New Haven Clock Co. in 1853 and served as president again making movements solely for Chauncey Jerome. After 1856 the New Haven Clock Co. purchased the bankrupt Jerome Manufacturing Co. and began making complete clocks with their own movements and cases. The question is, is there any way to tell an after-1856 movement from the New Haven Clock Co. from the earlier Chauncey Jerome movement? The answer is yes. See figure 9.2f.

clocks through the American Clock Co., a New York City sales agency formed in 1850 to sell Connecticut clocks. New Haven Clock Co. clocks were included in the American Clock Company's catalogs. In the 1870s they produced mainly OGs with some steeples, cottages, beehives, and lever wall clocks. This meant that the New Haven Clock Co. was largely a wholesaler of clocks to England through the American-owned sales firm Jerome & Co. of Liverpool and through the American Clock Co. of New York City.

In 1879, however, the New Haven Clock Co. published its first catalog. By 1890 business slowed and Hiram Camp resigned as president. The company was near bankruptcy in 1896 but survived. In 1900 business improved and was declared the best in 20 years.

The twentieth century is beyond the scope of this book but, briefly, New Haven continued making clocks until the Second World War, reaching three million clocks a year. Production then switched to war-related items. After the war the name changed to the New Haven Clock & Watch Co. In 1956 the company became bankrupt. From March 22 to 24 of 1960 all property, including the factories, was sold for around $500,000.

The New Haven Clock Co. produced a full line of American clocks in keeping with the times. They were a middle-of-the-road company making good standard clocks, but never distinguishing themselves. I consider the OG as their mainstay clock, especially the OG clock produced with the Jerome & Co., New Haven, label for sale in Britain, figures 9.2a-9.2e.

Figure 9.2f. This is a Chauncey Jerome movement dating from the late 1840s. Notice the shape of the escape wheel cock. It is stocky and has parallel sides over the teeth of the escape wheel. In figure 9.2e you can see that this later bridge is longer and tapered over the escape wheel teeth. That is the major difference between the otherwise nearly identical movements. Escape wheel bridge stocky with parallel sides—early; slim with tapered sides—later. This only applies to New Haven and Jerome movements because other Connecticut manufacturers used escape wheel bridges with similar shapes but they have no significance as to age.

INGRAHAM

Elias Ingraham, born in 1805, was trained as a cabinetmaker. It is ironic that the first three firms discussed in this chapter—Seth Thomas, New Haven and Ingraham—were formed by men trained to work with wood. Seth Thomas was a joiner and Chauncey Jerome and Elias Ingraham were cabinetmakers. Only Hiram Camp, the long-time president of New Haven, was trained as a maker of metal clock movements.

Ingraham came to Bristol in 1828 at the invitation of George Mitchell to devise new designs for clock cases and make them for Mitchell. He was a success, at least at designing and making the cases. He was not always successful at business. In 1835 Ingraham was still in Bristol working independently, making clock case parts for Davis & Barber of Macon, Georgia. He finished that contract in 1836 and turned to manufacturing a patented "premium rocking chair and reclining chair." This proved successful until he was caught up in the Panic of 1837. Weakened financially, he survived until 1840 when he petitioned for bankruptcy.

Elias Ingraham's brother Andrew purchased Elias's factory at the trustees' auction and sold a half interest to Benjamin Ray, forming the clock case manufacturing firm Ray & Ingraham. They hired Elias Ingraham as an employee. Though successful, the firm was dissolved in 1843

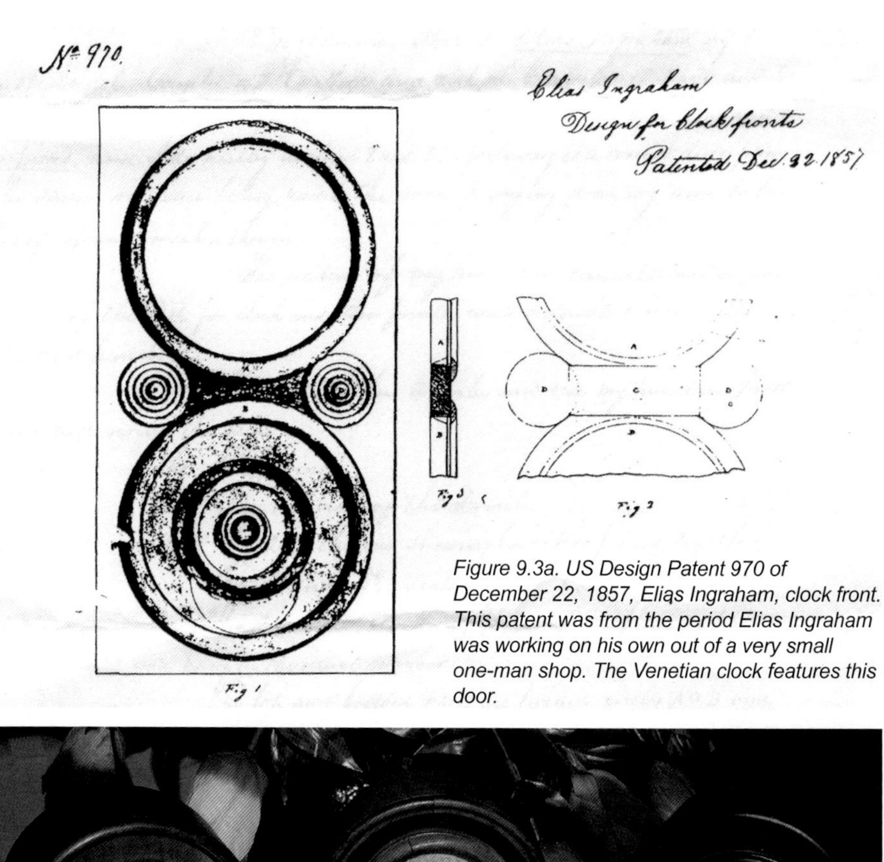

Figure 9.3a. US Design Patent 970 of December 22, 1857, Elias Ingraham, clock front. This patent was from the period Elias Ingraham was working on his own out of a very small one-man shop. The Venetian clock features this door.

Figure 9.3b. Three 13 inch Ingraham Venetian clocks from the 1870s. The clocks on the left and right are both gilt column models while the clock in the center has a raised arch molding that extends around the front of the clock. All three clocks use the Ingraham patent door.

Figure 9.3c. The clock in the center of figure 9.3b was a 30-hour time-and-alarm clock but at present has no movement. The green E. Ingraham & Co. label can be seen through the open lower door.

Figure 9.3d. The Venetian came in five sizes: 20 inch, 18 inch, 15 inch, 13 inch, and 12 inch. These three clocks are 18 inch, 15 inch, and 13 inch.

Figure 9.3e (opposite). The eight-day spring-driven movement from the 18 inch Venetian. The movement is stamped "E. Ingraham & Co / Bristol, Conn." It has an escape wheel bridge similar to the Jerome OG clock bridge, but that is only a coincidence and is not significant. Thanks to Bill & Rusty Bergman of Dayton, Ohio, for allowing me to photograph their clocks in their home.

after producing 10,000 clock cases, mostly OGs.

In late 1843 brothers Andrew and Elias Ingraham formed a partnership with the movement makers Elisha C. Brewster. The firm, Brewster and Ingrahams, was instantly successful. Elias Ingraham had designed the beehive, steeple, OG round Gothic or onion top, and the four-column steeple cases. Elias Brewster had developed a stable of successful eight-day and 30-hour brass spring clock movements with and without fusees. The combination proved explosive—and explode on the market they did. Within 18 months of its formation, Brewster & Ingrahams was the largest manufacturer of clocks in the world, helped along by the fact that Chauncey Jerome lost his Bristol factories to fire in April of 1845. Brewster & Ingrahams had also hired Jerome's senior sales agent in England, Epaphroditus Peck, away from Jerome in 1844.

Within a couple of years Jerome had recovered from his fire disaster, moved to New Haven, Connecticut, and regained his number one ranking. Brewster & Ingrahams was still very successful until the partners had a squabble, again over English sales agents, one of whom was E. C. Brewster's son. This led to the break-up of Brewster & Ingrahams in July of 1852. The brothers Elias & Andrew Ingraham carried on as E. & A. Ingrahams & Co.

Disaster struck in 1855 when fire destroyed the Ingraham's Bristol factory and in 1856 the

company went bankrupt.

Down on his luck—but not out—Elias Ingraham formed the very small firm of Elias Ingraham & Co. in 1857 to make a new series of shelf clocks he designed, which he sold using movements he purchased from others. This time Elias Ingraham obtained design patents for his new cases, figure 9.3a. Encouraged by the success of the new Victorian Gothic revival designs, the firm of E. Ingraham & Co. was formed. Elias was now working with his son Edward but not his brother Andrew. This is the company that became one of the Connecticut "Big Seven" clock firms and E. Ingraham & Co. with minor name changes lasted until 1976.

The basic product line of the 1860s consisted of the designs of Elias Ingraham—the Venetian, Doric, Ionic, and Grecian shelf clocks—and these models, with minor variations, were still the mainstay of the company until 1880. In 1881 they introduced walnut parlor clocks, black enamel cases, nickel-plated alarm clocks, and two models of carriage clocks. Slowly over the years new models were produced and the designs of Elias Ingraham eliminated. Elias Ingraham died in 1885 and his son Edward took over the company only to pass away himself in 1892. In turn, Edward's son Walter assumed the presidency and the family stayed in control of the firm until 1957.

I have selected the Ingraham Venetian as the signature clock of E. Ingraham & Co., figures 9.3b-9.3e.

ANSONIA

Clock movements are made from iron and brass. In the accounting year ending June 1, 1850, the Bristol, Connecticut, clockmaking firm of Terry and Andrews used 12 tons of iron wire and 58 tons of brass. That was the greatest amount of brass of any of the 11 Bristol clockmaking firms. Probably Seth Thomas of Plymouth, Connecticut, and Chauncey Jerome of New Haven, Connecticut, used more brass, but those statistics are not available.

At this time the wealthy brass manufacturer, Anson G. Phelps, approached Theodore Terry and Franklin Andrews of Terry & Andrews and persuaded them to leave Bristol and co-locate with his brass operations in the village of Ansonia, Connecticut, which Phelps had named after himself. The new clock company was formed on May 7, 1850, as

Figure 9.4a. The Regulator "A." The clock is not a true regulator having no maintaining power and no second hand. This eight-day calendar model does, however, have a movement with a deadbeat escapement. One other model that does show seconds has the same higher-grade movement. This clock is 32 inches tall.

Figure 9.4b. The12 inch dial and hands are all original. This eight-day striking calendar model sold for $11.30 in 1901. Probably more of these clocks were sold in Britain than America, and the Ansonia sister firm, the British United Clock Co., sold a very similar clock.

the Ansonia Clock Company, again named after Anson Phelps. Theodore Terry was elected president. Phelps held 1,334 shares of stock while Terry and Andrews held 1,333 shares each.

In an earlier article we commented upon the fact that some of the great Connecticut clock firms were named after people, such as Seth Thomas, and others after places, such as the New Haven Clock Co. The firm of Ansonia was named after a man, Anson Phelps, *and* a place, Ansonia, which was also named after Anson Phelps.

At the start many of the Ansonia clocks made in 1851 were a combination of Terry and Andrews and Ansonia Clock Co. parts. A clock might have a dial signed "Ansonia Clock Co., Ansonia, Conn." and a movement stamped "Terry & Andrews, Bristol, Conn." They even printed labels as "Terry & Andrews, Ansonia, Conn." in spite of the fact there was never such a firm at that location.

Near the end of 1851 Franklin Andrews sold all but one of his shares. Less than two years

Figure 9.4c. The original Ansonia symbol on the dial.

Figure 9.4d (opposite). The deadbeat eight-day calendar movement is stamped "Ansonia Clock Co. USA New York." The clock was actually made in Brooklyn, a borough of New York City. The "USA" indicates the clock was made for sale overseas.

later Anson Phelps sold his shares in the clock company just prior to his death to his son-in-law James B. Stokes. The firm Phelps, Dodge & Co. took over the Ansonia Brass and Copper Company and ended up with their 2,000 shares of the Ansonia Clock Co. while Theodore Terry held 1,999 shares and his son Hubbel P. Terry held one. All this fighting for control of the clock company was for naught and in late 1854, just when the firm was doing well after relocating from Bristol to Ansonia, the factory burned to the ground.

As an aside, the story of each of the clock firms always seem to include one or two fires that destroyed the firm's factory and machinery. Sometimes the firms bounced back but often they didn't. In this case they gave up and sold the "land & ruins" and later the balance of the firm's rights to make clocks etc. to Phelps, Dodge & Co. This ended the company, or at least put it to sleep for the next 15 years.

During that period from 1855 to 1870, I imagine Ansonia as a little kitten in a country village wandering here and there and not belonging to anyone, but surviving. Clock movements were made at Ansonia by Phelps, Dodge & Co. under various names such as Ansonia Brass & Battery Co. and Ansonia Brass Co. They were mostly supplying these movements to others.

The clock portion of the giant brass producer Phelps, Dodge & Co seems to have come back to life between 1869 and 1870 when, operating under the name Ansonia Brass & Copper Co, they produced 83,503 clocks (with cases) using 90,000lb of brass in the process. By 1870 they were offering 45 models of clocks and 14 movement models.

In 1877 that clock portion of Phelps, Dodge & Co was reorganized at New York City (Brooklyn) using the old name, Ansonia Clock Co. An important addition to the company was Henry J. Davies. Henry was one of three Davies brothers from England. All ended up as significant members of the Ansonia Clock Co. and one, Edward Davies, returned to England to form and manage the British United Clock Co., which retained informal connections with Ansonia and produced many very similar clocks.

Henry Davies had been working as a clockmaker in New York City and had been using many of the Ansonia Brass & Copper Co. movements in his clocks. Phelps, Dodge & Co. bought Henry Davies's small clock firm and made him manager of the new Ansonia Clock Co. with his brother Edward Davies as superintendent. A New York City clock factory was built in Brooklyn,

New York. Brooklyn is a borough of New York City and only about 25 miles from Connecticut. With the move nearly completed, in 1880 the factory was destroyed by fire.

The advantage of a clock company having a larger parent company, this time a major brass producer, is that capital was available to quickly rebuild after the fire. By 1886, after a very successful run of about 25 years, Ansonia had $600,000 in stock, $500,000 due them from clock sales, and no debt. At about this time, it seems, Henry Davies was no longer with the firm.

The Ansonia Clock Co. made quality clocks and a lower grade of watches. In addition to wooden clocks, they produced more metal and china clocks than the other Connecticut firms.

They were the only American firm that moved out of Connecticut in the nineteenth century, albeit only about 25 miles outside the state.

Ansonia never seemed to recover from the First World War. In 1920 they offered 136 clock models, down from 450 in 1914 and by 1927 that number dropped to 47. Strange as it may seem, the company sold all its equipment to the Soviet Union's Amtorg Trading Corporation and sent Ansonia workers to Russia for 18 months to get the factory up and running. It was still in full operation in 1956.

The clock I think represents Ansonia best is the Regulator "A" wall clock. I might have chosen a brass or china clock but felt the Regulator "A" name is synonymous with the company, figures 9.4a-9.4d.

WATERBURY

Just like Ansonia, the Waterbury Clock Company was formed to take advantage of its parent corporation's production of brass. Maybe it would be better to say that the Benedict & Burnham Manufacturing Company, a manufacturer of brass sheeting, formed the Waterbury Clock Co. to use its brass.

The Waterbury Clock Co. was formed on March 5, 1857. There were 18 purchasers of its stock, with Aaron Benedict and G. W. Burnham of Benedict & Burnham being the largest. Noble Jerome supervised the movement shop and Edward Church the case shop. As an unfortunate aside, Noble Jerome was killed in 1861 when he was hit on the head by a 100lb chunk falling off a three-story building. He was replaced as movement superintendent by Silas B. Terry.

During the Civil War the company expanded even though one third of the employees left to serve in the Army at the start of the war. Then disaster struck. Fire destroyed both of their case shops on December 9, 1864. Fortunately, they recovered.

By 1867 the Waterbury Clock Co., was selling clocks through the American Clock Co., the sales agent in New York City. The company's products took a respectable 10 percent or more of the space in the sales agent's catalog. In 1873 their first price list showed Waterbury to be on par with the rest of the Connecticut firms, just smaller than some. In 1881 their first complete catalog was 122 pages and that grew to 175 pages in 1891. All but a few imported French pinwheel regulators were of their own manufacture and even those French regulators had Waterbury cases. In 1887 their sales agent in Edinburgh,

Scotland, was Thomas R. Dennison. He used a complete overpaste label on several of the American clocks he sold, mostly OGs, and the label stated the firm was the Waterbury Clock Co. of Edinburgh, Scotland.

In 1880 their parent company, Benedict & Burnham, incorporated the Waterbury Watch Co. separate from the clock company. By 1889 the Waterbury Clock Co. introduced an inexpensive watch in direct competition with their parent company's watch company. This large "watch" used the "Wasp" clock movement, a small clock movement for little novelty clocks. The independent firm Ingersoll visited the Waterbury Clock Co. and in 1892 placed an order for 1,000 of their watches for 85 cents each. By 1895 the clock company was selling Ingersoll half a million watches a year and by 1898 over a million. By 1910 the number of watches produced by the clock company was 10,000 a day. During this time of robust watch sales, clock sales continued but were not a point of interest for the company.

The Waterbury Clock Company stayed in business through its watch sales and in 1932 changed its name to Ingersoll-Waterbury and during the Second World War changed it again to the United States Time Corporation. After the war they developed the Timex watch. In 1969 the firm became the Timex Corporation and plants were operating in the American South. By 1997 the Timex watch still held a third of the American watch market.

I have selected the Waterbury Augusta as typical of the high grade wall clocks produced by the company, figures 9.5a to 9.5c. This clock is 4 feet 3 inches tall, yet is quite broad with a 10 inch dial. It is probably the most beautiful of all the Waterbury wall clocks being produced at the end of the nineteeth century. This clock is illustrated in Waterbury's 1893 catalog and sold

Figure 9.5a (opposite). The Augusta. The Waterbury Augusta is a beautiful wall clock over 4 feet 3 inches tall. It was made around 1893. There were over 50 different models of these Waterbury wall clocks all named after cities. They are very much in demand today and this Augusta model, named for Augusta, Georgia, tops the list. Photograph courtesy of Greg McCreight.

Figure 9.5b. A closeup of the Waterbury Augusta with the 10 inch dial removed. Photograph courtesy of Greg McCreight.

Figure 9.5c. The eight-day chain-drive Waterbury Augusta movement is of good quality, but has no regulator features. Photograph courtesy of Greg McCreight.

for $30.

There is always a question whether these Waterbury wall clocks were domestic clocks or meant for the office. I think this very fancy clock would be at home in the home as well as in the office of a bank president. It is a weight-driven clock with no second hand. The movement is not high grade and has no regulator features.

Gilbert

Born in 1806, William L. Gilbert lived to be 84 years old, and 62 years of his life were spent in the clock business. We have talked before of the background American clockmakers had prior to entering the clock trade. Chauncey Jerome, Elias Ingraham, and Seth Thomas were cabinetmakers or joiners. William L. Gilbert, on the other hand, was a schoolteacher who had taught school for one year, 1827, and was not asked to return for the next. He was also the

Figure 9.6a. The Altai. This model is illustrated in Gilbert's 1885 catalog. It is a slant-sided walnut parlor clock of a style known as "Eastlake" in America and is over 20 inches tall. The clock's pendulum was patented by the company and has a moving hand on its pendulum dial that is supposed to aid in regulating the clock. Photograph courtesy of David and Brent Cox.

brother-in-law of George Marsh who had worked for Samuel Terry at Plymouth, Connecticut in 1827.

In 1828 Marsh and Gilbert purchased an old clock factory in Bristol for $900 and the firm Marsh, Gilbert & Co, was formed.

By 1832 both men were in nearby Farmington, Connecticut, and began producing clocks there, again as Marsh, Gilbert & Co., but sometimes with labels as George Marsh & Co. or just George Marsh. Marsh left Connecticut for Ohio and arrived in Piqua, Ohio, by 1834, probably selling clocks along the way. Gilbert was probably shipping Marsh clocks from Connecticut and there are clocks with a label just "George Marsh" with no place which were likely meant for sale in Ohio.

In 1835 Gilbert went into partnership with John Birge as Birge & Gilbert until 1837 when the Panic of 1837 shut down the company. After the nation's recovery, Gilbert was involved with a Bristol satinette and woolen company in 1839. That firm went under in 1840. During 1839 the "famous" clockmaking firm of Jeromes, Gilbert, Grant & Co., was formed to manufacture the first OG clocks using Noble Jerome's cheaper 30-hour brass movement. It was a huge success and the Jeromes bought out the other partners in 1840 for a 100 percent profit. Gilbert's share was thought to be $6,000.

Chauncey Jerome, writing in 1860 of Gilbert at the time of the breakup of Jeromes, Gilbert, Grant & Co., said: "He is said to be miserly in feeling, and is quite rich; not very enterprising, but has made a great deal of money by availing himself of the improvements of others."

Gilbert moved to Winchester, Connecticut, in 1840. There he, along with Lucius Clark and

Figure 9.6b. The pendulum of the Altai showing the dial and hand to assist in regulating the clock. Photograph courtesy of David and Brent Cox.

Figure 9.6c (above right). The Altai eight-day movement has a patent date of June 3, 1879. There is a Geneva stop on each winding arbor to prevent overwinding. Photograph courtesy of David and Brent Cox.

Ezra Baldwin, purchased the old clock factory of the late Riley Whiting for $5,000. The firm of Clark, Gilbert & Co. began making the cheaper brass clocks there in OG cases. By 1845 they were producing 12,000 clocks a year, not a large number by Bristol standards in 1845. At the end of 1845 Clark and Baldwin sold out to Gilbert for $4,666 and Gilbert renamed the firm Wm. L. Gilbert & Co. It is this firm that was to become one of the "Connecticut Seven." It almost didn't, for in 1848 Clark repurchased shares in the company for $5000 and it became Gilbert & Clark until 1851 when Clark again sold out to Gilbert and once again it was Wm. L. Gilbert & Co. By the way, the "& Co." was Gilbert's brother-in-law Isaac B. Woodruff.

The firm of Wm. L. Gilbert & Co. was one of the few that survived the mid-1850s depression. Clockmaker Silas B. Terry was not so lucky and declared his firm bankrupt in 1859. Gilbert hired Terry as his general manager in 1860.

The Civil War, 1861-1865, brought industrial gains to the North, including Connecticut. After the war people were ready to purchase consumer products, including clocks. To raise capital to expand his clock business, Gilbert formed the Gilbert Manufacturing Co. in 1866 and sold $4,000 of capital stock, 480 shares, to himself, 480 shares to George B. Owen, 520 shares to Isaac Woodruff, and 20 shares to N. S. Pond. George B. Owen, an Englishman, was chosen president until 1871 when Gilbert replaced him. Owen stayed on as general manager, replacing Terry.

Like all the successful Connecticut clock firms, Gilbert used the services of the New York City sales agent, the American Clock Company, to market its clocks in the late 1860s and early 1870s.

A fire destroyed the Gilbert Manufacturing Co.'s factory in 1871, but the company survived. A new factory was built and production commenced in 1873 under a new name, the William L. Gilbert Clock Co. The first of this company's clocks made it to the market in very early 1875. In that year Gilbert issued its first catalog.

The last quarter of the nineteenth century was good for Gilbert. Under the management of George B. Owen from 1866 to 1913 it prospered. Owen was granted 22 patents along the way. In 1913, at age 80, he was replaced by his son Arthur W. Owen. It is interesting that during the time George B. Owen was general manager of the William L. Gilbert Clock Co., he had his own clock company, George B. Owen, where he made the first walnut parlor clocks in America in 1875. His factory also burned, in 1879, but he built a new factory and called it the Winstead Clock Co. which he sold to the William L. Gilbert Clock Co. in 1894. It was closed in 1895, reopened in 1898, sold by Gilbert in 1899, sold back to Gilbert in 1903, burned down in 1903, rebuilt by Gilbert in 1903, sold back to George B. Owen in 1903 and the building still survives 108 years later.

William L. Gilbert died in 1890 at age 84. In death he was very generous giving money to many worthwhile projects and institutions. One example is a gift of a $50,000 endowment to found and maintain the Gilbert Seminary at Winstead, Louisiana, for African American children. He had no children of his own. After his death the presidency of the firm fell to his brother-in-law, Isaac B. Woodruff and thereafter to Isaac's son, James Gilbert Woodruff.

Back to the William L. Gilbert Clock Co. In 1887 Gilbert furnished the Davies Clock Co. of Columbus, Mississippi, with 8,000 OG weight clocks for $1.85 per clock. These were sold in the South with Davies Clock Co. labels. Gilbert spring OG and calendar OG clocks were also sold with various Southern labels.

By 1907 the Gilbert firm had sales offices in New York, Chicago, San Francisco, Philadelphia, and St. Louis. By 1930 they had a line of electric clocks on the market. The production of both wooden-cased clocks and eight-day brass spring movements ceased in 1940. They made Bakelite cases during the war years as well as molded papier-maché alarm clocks. After the war the firm had hard times and was sold, renamed, diversified until 1964 when all the clock production and the facilities were sold to the Sparta Corp., of Chicago, Illinois, and Louisville, Mississippi.

The Gilbert clock I have chosen to represent the company is the Altai, a walnut shelf clock introduced in 1885. George B. Owen was managing Gilbert at that time and, as we noted, he had simultaneously started his own independent clock company, George B. Owen, and introduced the walnut parlor clock to America in 1875. The Altai was a Gilbert clock but typical of the successful George B. Owen walnut parlor clocks, figures 9.6a-9.6c.

Welch/Sessions

The companies of Welch and Sessions can be thought of as one company. Over a period of time the Sessions family gained control over the firm and on January 9, 1903, simply changed the name from the E. N. Welch Manufacturing Co. to the Sessions Clock Co. The only difference between the two companies was the name. No ownership or management changes. No change in the clocks produced at that time.

The company of E. N. Welch can be traced back to the Forestville Manufacturing Company and J. C. Brown. Forestville is a district in the town of Bristol, Connecticut. In 1835 a clock shop was located in Forestville owned by Jonathan C. Brown, William Hills, Lora Wells, Chauncey Pomeroy and Jared Goodrich. The firm was called the Forestville Manufacturing Co. and J. C. Brown was the principal owner.

Figure 9.7a. The Patti VP. Named for opera singer Adelina Patti (1843-1919), this is the full-size 19 inch tall Patti with an eight-day movement striking on a bell. VP simply stands for "Visual Pendulum." Photograph courtesy of David and Brent Cox.

Figure 9.7b. The Patti No. 2. Commonly referred to as the "Baby Patti," standing only 11 inches tall. This example is eight-day time-only and is a hard clock to find. Photograph courtesy of David and Brent Cox.

The firm declared bankruptcy during the Panic of 1837 and was forced to cease business in 1840. Somehow, J. C. Brown was able to keep control of the factory and continued the operation on a shoestring from 1840 to 1856 under various names and with many combinations of partners. Labels were sometimes J. C. Brown and sometimes Forestville Mfg. Co., or sometimes both on the same clock.

Faced with bad economic times, J. C. Brown was bankrupt in 1856. Several other local businessmen had also gone under. Local financier Elisha N. Welch came forward and purchased J. C. Brown's Forestville Manufacturing Co., the Forestville Hardware and Clock Co., and casemaker F. S. Otis's business. He successfully molded these three firms into a profitable clockmaking concern and on July 6, 1864, formed the E. N. Welch Manufacturing Company. The E. N. Welch Mfg. Co. made quality products and were the largest producer of clocks in greater Bristol from 1865 to 1885.

At the same time that Elisha Welch was operating the E. N. Welch Mfg. Co. (1868-1884) he was involved in a partnership with Solomon C Springs in the successful production of high quality clocks under the name Welch, Spring & Co. in Bristol. Welch produced the movements and Spring the cases. The cases were noteworthy, made of the finest rosewood veneers. Once one has seen a Welch, Spring & Co. rosewood case, it is not difficult to identify another by the

Figure 9.7c. The Gerster VP. Note the "porch railing" top. Like the Patti VP, this clock stands 19 inches tall. Like Adelina Patti, Gerster was an opera singer. Photograph courtesy of David and Brent Cox.

Figure 9.7d. The back of the Gerster showing the remains of the Gerster label. Photograph courtesy of David and Brent Cox.

quality of the case and woods alone.

In 1877 the cost of Welch, Spring & Co. clocks was from $4.75 to $40 while E. N. Welch Mfg. Co. clocks were priced from $1.80 to $13. In 1884 Welch, Spring & Co. went out of business and their clocks were incorporated into the E. N. Welch Mfg. Co. product line. Sometimes the high-grade rosewood veneers were replaced with cheaper walnut on the E. N. Welch models.

With the death of E. N. Welch in 1887, the clock company went into a decline and by 1893 they literally shut down the business, let the workers go, and sold clocks they had already manufactured to pay down debt. The company stayed shut down from 1893 to 1897. Not a typical approach by any means, but it worked. By late 1897 the firm was back in operation and out of debt. Then on March 17, 1899, the inevitable happened. Fire destroyed half a dozen buildings including all the movement making capability. Rebuilding took place quickly and in six months they were back in full operation, when, in December 1899, another fire destroyed the case shop. Three months later a new case shop was in production.

Figure 9.7e. The Cary VP. Note the "sausage" top. This is slightly taller than the Patti and the Gerster at 20 inches. Again Cary was an opera singer. Photograph courtesy of David and Brent Cox.

Figure 9.7f. Typical Patti movement. The same movement is used in all of these clocks except the "Baby Patti," which has a scaled-down version of this movement. These movements have X-shaped plates, a club-tooth escape wheel and double mainsprings with single wind. Each great wheel arbor has two mainsprings stacked one on the other. The pendulums for these clocks are Sandwich glass. The Sandwich glass factory of New England produced "fancy" glass in the nineteenth century. Photograph courtesy of David and Brent Cox.

Even with a totally new factory with modern equipment, they were never able to recover from the debt they incurred, and by 1902 it was evident they were going under. This is when the Sessions family came to the rescue. In June of 1902 the company voted "that the affairs and business of this company be wound up and it be dissolved." Three members of the Sessions family bought the company stock: Albert L., William Edwin, and John Humphrey Sessions. Together, they owned 4,000 of 4,000 shares of preferred stock, and 11,960 of 12,000 shares of common stock. On January 3, 1903, the three Sessions family members voted to change the name to the Sessions Clock Co. Quickly, with an infusion of Sessions' cash, the firm came back to life.

Sessions' product line was much like the other Connecticut firms, but the company never attempted to produce prestige models. They introduced electric clocks in 1929 and in 1942

clock production was stopped for the war and resumed in 1945. The workers struck in May of 1967 and the firm never recovered. On October 1, 1968, they stopped production of clocks.

There is one line of clocks produced by Welch that they are most known for. The clocks are named after opera singers, the most famous of whom was Adelina Patti. In addition to the Patti model, there were the Verdi, Wagner, Cary, Gerster, and Scalchi clocks, figures 9.7a-9.7g.

I will close with a comment about these later mass-produced clocks that collectors should be aware of. One interesting thing about collecting mass-produced clocks is that serious collectors demand that everything be original whenever possible. It is a lot like collecting stamps or coins, also mass-produced items. Perhaps a somewhat rare stamp with a small imperfection is all that one can afford. When one buys a stamp with an imperfection one does not expect to go out and restore it: a "restored" stamp would be completely worthless. On the other hand, a one-off hand-made clock such as a longcase can be altered slightly by restoring it and nobody will ever know exactly what the original actually looked like and the value may even go up with the restoration. Buy the best example of a mass-produced clock that you can find or afford. Once you have it, try to keep the outside look of the clock just as you found it. You can lightly clean it, but don't go painting dials or "dressing it up." It is not money well spent.

Figure 9.7g. Patti No. 7 wall clock. This clock is 4 feet tall. Note the similarity between this clock and the Patti, figures 9.7a and 9.7b. This is a very rare clock, an early wall clock version of the later Patti No. 7 Regulator. The later regulator is a true regulator with maintaining power, weight drive, etc. This spring clock is so rare I had never seen one nor even heard of one prior to receiving this photograph. The spring wall clock was produced by Welch, Spring & Co., starting in 1882 prior to the true regulator. Note the gold leaf pendulum rod. All of the Patti book clocks were first produced by Welch, Spring & Co. but were produced with E. N. Welch Mfg. Co. labels after 1884. Photograph courtesy of David and Brent Cox.

APPENDIX 1
How Clocks Work

Most clocks are comprised of a case and a movement, the case being the (usually wooden) structure in which the (usually metal) movement is held. While the case is normally purely decorative, the movement is the mechanism which makes the clock work.

There are four principal components to the clock movement; the source of power (usually a weight or a spring); a "train" of wheels (or gearbox); an "escapement"; and a small system of gears, the "motionwork", which carries the hands.

The sentence I have just finished writing illustrates the difficulty of talking about clocks without using technical terms. To the horological beginner, it may seem as if he or she has to master a new language at the same time as trying to learn about clocks. For this reason I have included a full glossary of horological terms as one of the appendices to this book (page 125).

The average person will see a clock's movement as a mass of gears contained between two

Figure A.1. Movement of an American clock, showing the front plate (rectangular) with the escape wheel (top center) and two trains of wheels, the striking train (left) and the time train (right).

Figure A.2 (left). Clock wheel on its arbor with integral cut pinion.

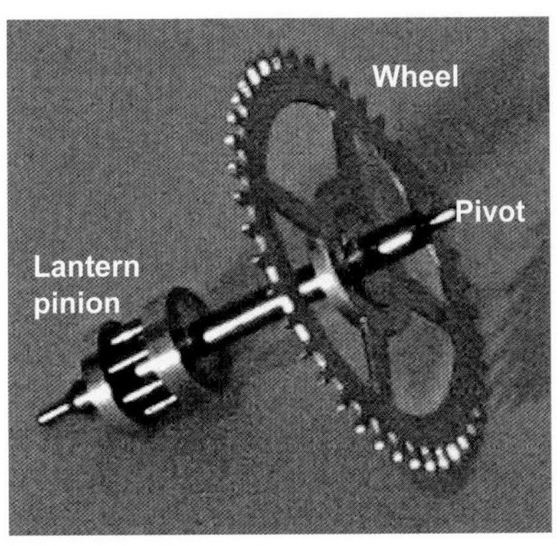

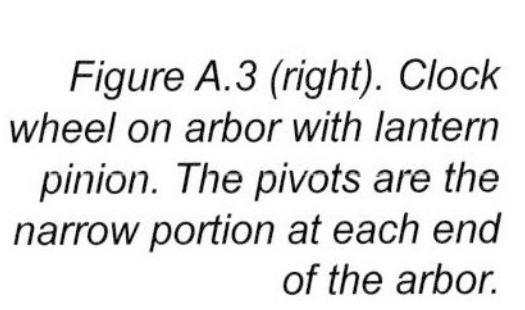

Figure A.3 (right). Clock wheel on arbor with lantern pinion. The pivots are the narrow portion at each end of the arbor.

flat sheets of brass, or "plates." To the horologist, the gears are "wheels" and "pinions," which normally exist in pairs.

A "wheel" is the larger gear of the pair, the pinion being the smaller one. The wheel is the driving element while the pinion is driven—except in some portion of the motionworks described later.

Pinions can be of two types, "cut" and "lantern." Later mass-produced brass-movement American clocks normally have lantern pinions.

The wheel-and-pinion pair is mounted on a shaft called an "arbor." At its ends, the arbor is reduced in diameter and those reduced portions of the arbor are "pivots." These pivots run in holes in the plates just slightly larger than the pivots themselves, allowing the wheel assembly to rotate freely.

For convenience, clock wheels are numbered from the "first wheel," to which the spring is attached, through the "second wheel," "third wheel" and so on. A series of wheels such as this is called a "wheel train." There may be several such wheel trains in a clock movement, notably the "time train" (which runs the hands) and the "strike train" (which powers the hammer which strikes the bell or gong). Confusingly, numbered train wheels may also be named. The first wheel may also be known as the "great wheel," while the last wheel of the time train is always the "escape wheel."

A clock doesn't really run, it starts and stops. The "tick-tock" you hear is the clock continually stopping—it doesn't make any noise you can hear upon starting. The part of the clock that controls this starting and stopping is the "escapement," which is one of the best terms ever used to describe anything. The escapement allows the power in a clock to escape for an instant, starting the train turning, and then stopping it.

The parts of the escapement are the "escape wheel" and, in its most common form, an "anchor" or "verge." The latter is normally a bent strip of steel with, at each end, a flat face ground into it. The two flat faces are known as the "pallets." Each face is a hard, polished metal surface and it is the noise of a tooth coming to a halt against it which causes the clock's characteristic "tick."

The pendulum controls the rate at which a clock runs. The shorter the pendulum, the faster it goes. To allow the pendulum to swing freely back and forth there is a short section of flat metal at the top called the "suspension spring." At the bottom of the pendulum is a piece of metal, usually a disc, called the "bob." A screwed nut beneath the bob can be used to adjust it upwards to go faster or downwards to go slower.

Most clocks have a center arbor that rotates once an hour which is extended through the front

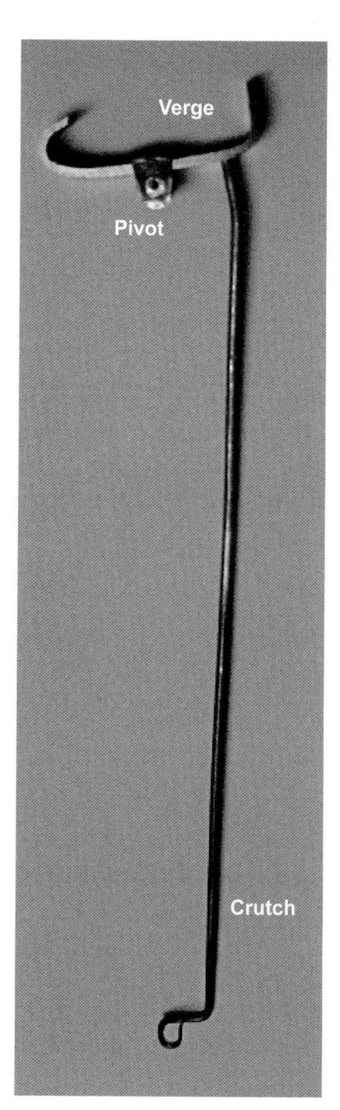

Figure A.4. The verge (top) and attached pendulum crutch. This assembly rocks to and fro on its pivot, allowing the escape wheel to move around one tooth at a time.

plate, then through a hole in the face of the clock, or "dial." The minute hand attaches to this arbor. Behind the dial the motionworks reduces the rate of revolution by 12 : 1 and drives the hour wheel which is mounted over the center arbor and also extended through the dial and carries the hour hand.

Most clocks strike and there are two common methods of controlling striking, the "rack-and-snail" and the "count wheel." Both rely upon striking a sequence of one to 12 strikes upon a bell or gong over a 12-hour period and then repeating the sequence over the next 12 hours.

A clock is wound with a "key" or if it has weights, a "crank" which can be turned continuously without letting go to get a new grip.

APPENDIX 2
FURTHER READING

1. *Clockmakers & Watchmakers of America by Name and by Place*, 2011 (2nd edition), by Tom & Sonya Spittler & Chris H. Bailey, National Association of Watch & Clock Collectors, Columbia, Pennsylvania. Website: www.nawccstore.org.

2. *American Wooden Movement Tall Clocks 1712-1835*, 2011, by Philip E. Morris Jr., Heritage Park Publishing, Hoover, Alabama. Website: www.heritageparkpublishing.com.

3. *Eli Terry and the Connecticut Shelf Clock*, 1994 (2nd edition) by Kenneth D. Roberts and Snowden Taylor. Self-published and available from Snowden Taylor, 318 Western Highway, Tappan, NY10983, as is companion book (4) below.

4. *The Contributions of Joseph Ives to Connecticut Clock Technology 1810-1862*, 1988 (2nd edition) by Kenneth D. Roberts, see (3) above.

5. *Willard's Patent Timepieces* (2002) by Paul J. Foley, Roxbury Village Publishing of Roxbury, Massachussetts. Website: www.roxbury-village-publishing.com.

6. *Ansonia Clocks & Watches*, 1998 (2nd edition) by Tran Duy Ly, Arlington Books, 215 Shadowood Drive, Johnson City, Tennessee 37604-1128. Website: www.arlingtonbooks.com.

7. *Gilbert Clocks*, 1998, (with supplement) by Tran Duy Ly, Arlington Books, as above.

8. *Ingraham Clocks & Watches*, 1998, by Tran Duy Ly, Arlington Books, as above.

9. *Kroeber Clocks*, 2006, by Tran Duy Ly, Arlington Books, as above.

10. *New Haven Clocks & Watches*, 1997, by Tran Duy Ly, Arlington Books, as above.

11. *Sessions Clocks*, 2001, by Tran Duy Ly, Arlington Books, as above.

12. *Seth Thomas Clocks & Movements* (2 volumes), 2004 (3rd edition), by Tran Duy Ly, Arlington Books, as above.

13. *Waterbury Clocks & Watches*, 2001 (2nd edition) by Tran Duy Ly, Arlington Books, as above.

14. *Horology Americana*, 1972, by Lester Dworetsky & Robert Dickstein, Horology Americana Publishing, New York, NY.

15. *The American Clock*, 1976, by William H. Diston & Robert Bishop, E. P. Dutton & Co., Inc., New York, NY.

16. *The Longcase Clock Reference Book* (2 volumes), 2001, by John Robey, Mayfield Books, Ashbourne, Derbyshire, UK. Website: www.mayfieldbooks.co.uk.

17. *Two Hundred Years of American Clocks and Watches*, 1975, by Chris Bailey, Prentice Hall Publishing, New Jersey.

APPENDIX 3
GLOSSARY

Arbor Spindle, shaft, or axle on which clock *wheels*, *pinions,* etc. are mounted.
Back cock Single-foot bracket, usually of brass, that holds the rear *pivot* of the *escape wheel.*
Back plate *see* "Plates."
Banjo clock Clock said to resemble a banjo, invented in 1803 by Simon Willard, 1753-1848, of Roxbury, near Boston, Massachusetts.
Barrel A short cylinder of brass that contains the *mainspring.*
Beat (setting) The "tick-tock" of the clock as the escapement operates. The "tick" should be of equal length to the "tock," beat setting being the process of making sure this happens, after which the clock is said to be "in beat."
Bob The part of the *pendulum* at the far end of the rod from the suspension. The bob accounts for most of the weight of the *pendulum.*
Cannon pinion *Pinion* that drives the *motionworks* of a clock, part of which takes the form of a hollow cylinder (thought to resemble a cannon) onto which the minute hand fits.
Center arbor *Arbor* of center wheel that revolves once every 12 hours to which the *cannon pinion* is attached.
Center wheel The wheel, usually the third wheel of the time train, on the arbor of which the minute hand is mounted.
Chain, fusee Fine bicycle-type chain that winds onto a *fusee*, allowing the power of the mainspring to be delivered evenly to the *train.*
Chapter ring The ring on a clock *dial* that shows the hours or "chapters," showing religious derivation: the first clocks were in monasteries, reminding the various chapters of monks of the times of their religious duties.
Chime Short tune played by clock on the hours and quarters.
Click, clickspring The click is a ratchet pawl used in winding springs, the clickspring being a small spring that keeps the click in engagement with the ratchet.
Collet Circular ring used to hold clock parts on *arbors* etc.
Countwheel *Strike train wheel* which determines the number of times the clock strikes at a given hour.
Crutch Lever that transmits impulse from *escape wheel* to *pendulum* and transmits the beat from the *pendulum* to the *pallets.*
Dial Face of clock or watch against which hands show the time.
Escapement The mechanism at the end of the wheel *train* that allows the power to escape in regular bursts, thus allowing the accurate display of the passage of time.

Escape wheel Wheel at top of the time *train* that allows power to escape in tiny regular bursts, giving the clock its characteristic "tick-tock."

Escapement, bent-strip Escapement, the *verge* of which is fashioned out of a strip of metal bent to shape.

Finial Decoration used to "top-off" part of the case of a clock. Finials are commonly found at the top corners of the case of a *tallcase clock.*

Fly *Arbor* carrying vanes, usually at the end of the *strike train*, which acts as an air brake, slowing down the rate at which the *train* runs when striking takes place.

Fusee Conical construction from which a chain or gutline is pulled, allowing the power of the mainspring to be released evenly.

Girandole Type of *banjo clock* pioneered by Lemuel Curtis thought to resemble a girandole.

Gong Bar or coil of bell metal on which clocks *chime* and/or *strike.*

Gut (line) Gut or plastic cord that performs the same function as the *fusee chain.*

Hood Top section of the case of a *tallcase* clock, usually removable, that surrounds the dial.

Hour wheel Part of the *motionworks* to which the hour hand is fitted.

Leaf (leaves) The name given by horologists to a tooth (teeth) of a *pinion.*

Mainspring Coiled flat spring that powers spring-driven clock *movements*, often contained in a *barrel.*

Maintaining power Power that keeps clock running while being wound.

Minute wheel 12:1 wheel-and-pinion arrangement that converts the hourly revolution of the *cannon pinion* (and minute hand) into the 12-hour revolution of the *hour wheel.*

Motionworks Gearing that converts the hourly rotation of the *cannon pinion* into 12-hour rotation of the *hour wheel.*

Movement The clock mechanism.

Ogee or OG clock *Shelf clock* in rectangular case with dial at top and reverse-painted tablet below, with S-shaped or "ogee" molding around the edge.

Pallet Flat face of verge that comes into contact with *escape wheel* teeth.

Pendulum Mechanism consisting of a *bob*, *rod,* and *suspension spring* that, swinging back and forth, is part of the *escapement* of many clocks.

Pillars Cylindrical posts that hold apart the *plates* of a clock *movement.*

Pinion Small clock gear, usually paired with a *wheel* in a wheel *train.*

Plates Parts of the movement between which the wheels run. The plates carry the pivot holes, which carry the pivots of the *arbors*, plus the *motionworks* and, on striking and chiming clocks, a variety of levers.

Rack Toothed lever used in rack striking for counting off the hours.

Regulator Clock designed to be accurate, with *maintaining power* and seconds hand.

Rod The part of the pendulum between the *bob* and the *suspension.*

Seat board Wooden board on which the movement of a *tall clock* sits.

Shelf clock Clock designed to sit on a shelf or mantel (as opposed to a floor-standing *tall clock*).

Snail Part of *rack* striking mechanism that allows the correct number of hours to be struck.

Strike The act of sounding the hours (and often quarters) on a bell or *gong.*

Suspension, suspension spring System by which the *pendulum* is suspended from the *back cock* by a spring or, occasionally, a silk thread or on a knife-edge.

Swan neck Decoration at the top center of the *hood* of some *tall clocks* thought to resemble the neck of a swan.

Tablet Glass panel on some clocks, often reverse-painted or mirrored.

Tall or tallcase clock Floor-standing, or grandfather clock

Train, going, striking, chiming Sequence of *wheels* and *pinions* that carry power from the weight or *mainspring* to the *escapement* (going train) or the strikework (striking train) or the chiming mechanism (chime train).

Verge The part of the *escapement* of a clock or watch that interferes with the movement of the *escape wheel*, allowing power to escape from the going train at a regular rate.

Wagon-spring clock Clock developed by Joseph Ives, which used a leaf spring for the motive force.

Wheels The large gears used in clockwork (c.f. *pinions*).

Winding square Square shape filed on the *arbor* of a clock onto which the winding key fits.

INDEX